D0252598

Wonderland

ALSO BY STEVEN JOHNSON

Interface Culture: How New Technology Transforms the Way We Create and Communicate

Emergence: The Connected Lives of Ants, Brains, Cities, and Software

Mind Wide Open: Your Brain and the Neuroscience of Everyday Life

Everything Bad Is Good for You: How Today's Popular Culture Is Actually Making Us Smarter

The Ghost Map: The Story of London's Most Terrifying Epidemic—and How It Changed Science, Cities, and the Modern World

The Invention of Air: A Story of Science, Faith, Revolution, and the Birth of America

Where Good Ideas Come From: The Natural History of Innovation

Future Perfect: The Case for Progress in a Networked Age

How We Got to Now: Six Innovations That Made the Modern World

Wonderland

HOW PLAY MADE THE MODERN WORLD

STEVEN JOHNSON

RIVERHEAD BOOKS
NEW YORK
2016

RIVERHEAD BOOKS
An imprint of Penguin Random House LLC
375 Hudson Street
New York, New York 10014

Library of Congress Cataloging-in-Publication Data

Names: Johnson, Steven.
Title: Wonderland : how play made the modern world / Steven Johnson.
Description: New York : Riverhead Books, 2016.
Identifiers: LCCN 2016035344 | ISBN 9780399184482 (hardback)
Subjects: LCSH: Amusements—History. | BISAC: HISTORY / World. | TECHNOLOGY & ENGINEERING / Inventions. | SCIENCE / Philosophy & Social Aspects.
Classification: LCC GV1200.J64 2016 | DDC 790.1—dc23
LC record available at https://lccn.loc.gov/2016035344
p. cm.

International edition ISBN: 9780735211919

Printed in the United States of America
1 3 5 7 9 10 8 6 4 2

BOOK DESIGN BY AMANDA DEWEY

For Eric

CONTENTS

INTRODUCTION *1*

1. FASHION AND SHOPPING *17*

2. MUSIC *65*

3. TASTE *109*

4. ILLUSION *147*

5. GAMES *187*

6. PUBLIC SPACE *233*

CONCLUSION *279*

Acknowledgments 285

Notes 287

Bibliography 299

Credits 307

Index 309

Those mechanical wonders which in one century enriched only the conjurer who used them, contributed in another to augment the wealth of the nation; and those automatic toys, which once amused the vulgar, are now employed in extending the power and promoting the civilization of our species. In whatever way, indeed, the power of genius may invent or combine, and to whatever low or even ludicrous purposes that invention or combination may be originally applied, society receives a gift which it can never lose; and though the value of the seed may not be at once recognized; and though it may lie long unproductive in the ungenial till of human knowledge, it will some time or other evolve its germ, and yield to mankind its natural and abundant harvest.

—David Brewster,
Letters on Natural Magic

Toys and games are the preludes to serious ideas.

—Charles Eames

Wonderland

Introduction

At Merlin's You Meet with Delight

In the early years of the Islamic Golden Age, around 760 CE, the new leader of the Abbasid Dynasty, Abu Ja'far al-Mansur, began scouting land on the eastern edge of Mesopotamia, looking to build a new capital city from scratch. He settled on a promising stretch of land that lay along a bend in the Tigris River, not far from the location of ancient Babylon. Inspired by his readings of Euclid, al-Mansur decreed that his engineers and planners should build a grand metropolis at the site, constructed as a nested series of concentric circles, each ringed with brick walls. The city was officially named Madinat al-Salam, Arabic for "city of peace," but in common parlance it retained the name of the smaller Persian settlement that predated al-Mansur's epic vision: Baghdad. Within a hundred years, Baghdad contained close to a million inhabitants, and it was, by many accounts, the most civilized urban environment on the planet. "Every household was plentifully supplied with water at all seasons by the numerous aqueducts which intersected the town," one contemporary observer wrote, "and the streets, gardens and parks were regularly swept and watered, and no refuse was allowed to remain within the

An illustration of a life-sized elephant clock, from al-Jazari's The Book of Knowledge and Ingenious Mechanisms

walls. An immense square in front of the imperial palace was used for reviews, military inspections, tournaments and races; at night the square and the streets were lighted by lamps."

More significant, though, than the elegance of Baghdad's broad avenues and lavish gardens was the scholarship sustained inside the Round City's walls. Al-Mansur founded a palace library to support scholars and funded the translation into Arabic of science, mathematics, and engineering texts originally written in the days of classical Greece—works by Plato, Aristotle, Ptolemy, Hippocrates, and Euclid—along with Hindu texts from India that contained important advances in trigonometry and astronomy. (These translations eventually turned out to be a kind of lifeboat for these ancient ideas, keeping them in circulation through the European Dark Ages.) A few decades later, under the leadership of al-Mansur's son, al-Manum, a new institution took root inside Baghdad's walls, a mix of library, scientific academy, and translation bureau. It became known as Bayt al-Hikma: the House of Wisdom. For three hundred years, it was the seat of Islamic scholarship, until the Mongols sacked Baghdad in the siege of 1258, destroying the books from the House of Wisdom by submerging them in the Tigris.

In the first years of the House of Wisdom, al-Manum commissioned three talented brothers, now known as the Banu Musa, to write a book describing classical engineering designs inherited from the Greeks. As the project evolved, the Banu Musa expanded their brief to include their own designs, showcasing the advances in mechanics and hydraulics that surrounded them in Baghdad's flourishing intellectual culture. The work they eventually published, *The Book of Ingenious Devices*, now reads like a prophesy of future engineering tools: crankshafts, twin-cylinder pumps with suction, conical valves employed as "in-line" components—mechanical parts centuries ahead of their time, all represented in detailed schematics. Two centuries later, the Banu Musa's work inspired an even more astonishing project, written and illustrated by the Islamic engineer al-Jazari, *The Book of the Knowl-*

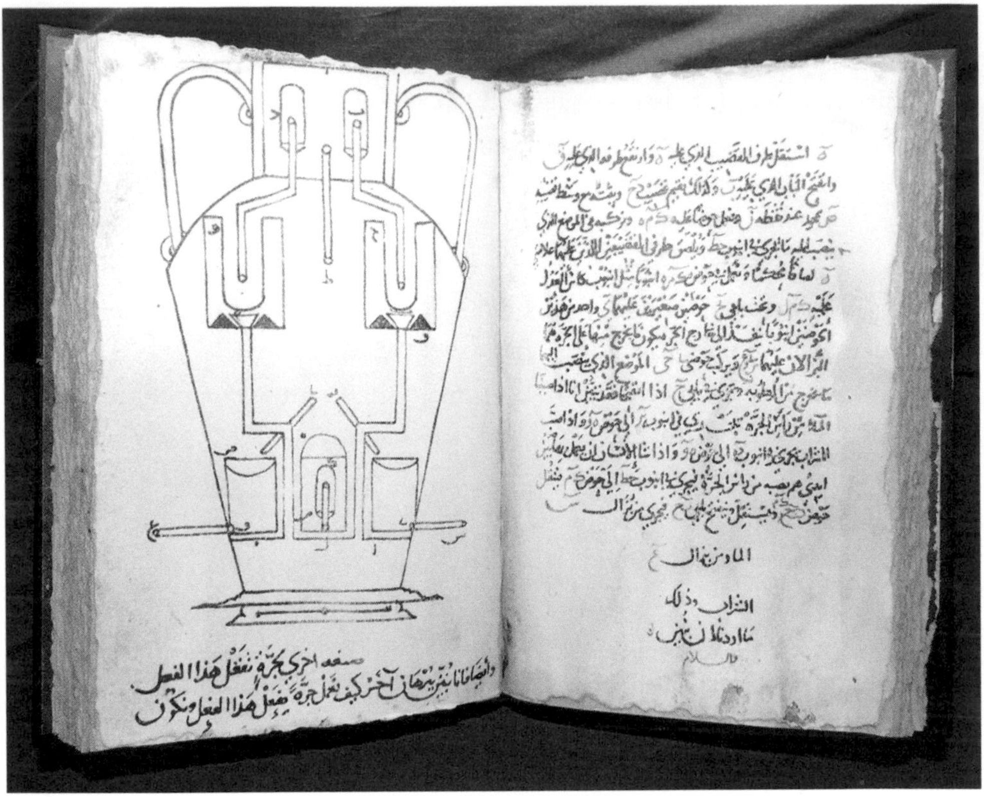

***Pages from the Banu Musa's* The Book of Ingenious Devices**

edge of Ingenious Mechanisms. It contained stunning illustrations, adorned with gold leaf, of hundreds of machines, with careful notes explaining their operational principles. Float valves that prefigure the design of modern toilets, flow regulators that would eventually be used in hydroelectric dams and internal combustion engines, water clocks more accurate than anything Europe would see for four hundred years. The two books contain some of the earliest sketches of technology that would become essential components in the industrial age, enabling everything from assembly-line robots to thermostats to steam engines to the control of jet airplanes.

These two books of "ingenious" machines deserve a prominent place in the canon of engineering history, in part as a corrective to the too-frequent assumption that Europeans single-handedly invented most modern technology. But there is something else about these two books that doesn't quite fit the standard account of groundbreaking scientific work, something that is immediately visible to the nonengineer flipping through their pages. The overwhelming majority of the mechanisms illustrated in the two volumes are objects of amusement and mimicry: fountains that spout water in rhythmic bursts; mechanical flute players; automated drumming machines; a peacock that dispenses water when you pull its feathers and then proffers a miniature servant with soap; a boat filled with robotic musicians that can serenade an audience while floating in a lake; a clock built into the shape of an elephant that chimes on the half hour.

There is a puzzle lurking in the genius of the Banu Musa and al-Jazari. How can it be that such advanced engineering expertise should be devoted to toys? The revolutionary ideas diagrammed in the pages of these ancient books would eventually transform the industrial world. But those ideas first came into being as playthings, as illusions, as magic.

Fast-forward a thousand years. The mechanical amusements first diagrammed by al-Jazari and the Banu Musa have become profitable entertainment across Europe, nowhere more so than in the streets of London, which teem with spectacles and curiosities. By the early 1800s, a bustling new industry of illusion has taken root in the West End. Robert Barker's immersive Panorama dazzles audiences with a simulated 360-degree rooftop view of the city; at the Lyceum Theatre, Paul de Philipsthal terrifies spectators with his multimedia spook show, the Phantasmagoria. An exhibit of wax statues, curated by a certain Madame Tussaud, premieres at the Lyceum, but isn't a hit. (Tussaud wouldn't create her famous museum for another thirty years.) In Hanover Square, just south of Oxford Street, a Swiss inventor and showman with the delightful name John-Joseph Merlin runs an eclectic establishment known as Merlin's Mechanical Museum. In modern terms, Merlin's shop is a kind of hybrid between a science museum, a gaming arcade, and a maker lab. You can marvel at moving mechanical dolls, try your luck at gambling machines, and enjoy the sweet melodies of music boxes. But Merlin is not simply an impresario; he is also a mentor of sorts, encouraging the "young amateurs of mechanism" to try their own hands at invention.

Born in Belgium in 1735, Merlin is a clockmaker by trade, and like many horologists of that period, he has long been intrigued by the idea that the mechanized movement of the pendulum clock and its descendants could be scaled up into more impressive feats—of productive labor, to be sure, but also something else: flights of fancy, wonder, illusion. You could build machines that could tell time, weave fabric, maybe even perform elementary calculations. But you could also build machines that mimicked physical behavior for less utilitarian purposes: for the sheer delight that human beings have always found in the imitation of life. The construction of these early robots, called automata in their time, had been one of the great extravagances of courtly life during the period, designed to amuse and curry favor with the aristocracy. These inventions evolved out of mechanical clocks, popular in the 1600s, featuring elaborate mise-en-scènes of villages or musicians that mark the passing of the hour by bursting into life. By the end of the seventeenth century, the clocks blossomed into miniature stage shows, called clockworks, that presented simple narratives using the mechanized movements of hundreds of distinct elements. Many of them featured biblical themes. In 1661, a London tavern showcased a clockwork rendition of Eden. According to a pamphlet published at the time, it presented "Paradise Translated and Restored, in a Most Artfull and Lively Representation of the Severall Creatures, Plants, Flowers, and Other Vegetables, in Their Full Growth, Shape, and Colour . . . A Representation of that Beautiful Prospect Adam had in Paradise." (When the robots eventually write the history of their species, these animated tableaux will serve nicely as a creation myth.)

By the early 1700s, the focus shifted from re-creating the bustle of an animated village or garden to building increasingly lifelike simulations of individual organisms. In the first half of the eighteenth century, the French inventor Jacques de Vaucanson famously constructed an automaton called the Digesting Duck that consumed grain, flapped its wings, and—the *pièce de résistance*—actually defecated after eating. A few decades later, in 1758, a Swiss horologist named Pierre Jaquet-Droz traveled to Madrid to present an array of wonders to King Ferdinand, most of them pendulum or water clocks that featured animated storks, flute-playing shepherds, and songbirds—the mechanical descendants of al-Jazari's ingenious devices. The audience with Ferdinand secured Jaquet-Droz financially and he embarked on an ambitious streak of automaton creation, arguably the most artistic and innovative mechanical engineering that the world had ever seen. His crowning achievement, completed in 1772, was the Writer, a mechanical boy composed of more than six thousand distinct parts, seated on a stool with a quill pen in hand. The boy could be programmed to write any combination of words using up to forty characters. Once instructed—via a series of cams hidden inside the contraption—he dipped his pen in an inkwell, shook it twice, and began writing the words with a studious precision, his eyes following the pen as he wrote. The Writer was not a computer in the modern sense of the word, but it is rightly considered a milestone in the history of programmable machines.

Jaquet-Droz's son, Henri-Louis, began displaying the Writer in London in 1776, part of a new exhibition in Covent Garden called the "Spectacle Mécanique." Inspired by these fantastical creatures, Merlin began making and collecting automatons himself. To show-

The Writer, an automaton created by Pierre Jaquet-Droz in 1772

case some of this work, he opened Merlin's Mechanical Museum in 1783, running a promotional notice assuring that "Ladies and Gentlemen who honour Mr Merlin with their Company may be accommodated with TEA and COFFEE at one Shilling each." As Simon Schaffer puts it, Merlin "prowled the borderlines of showmanship and engineering," not unlike the Hollywood special-effects studios that descend, almost directly, from Merlin and his contemporaries.

Merlin's ingenuity took him in many directions: he invented a self-propelling wheelchair, a mechanical Dutch oven, a pump that automatically freshens air in hospital rooms, a deck of playing cards with braille-like encodings that enables blind people to play whist. He dabbled in the design of musical instruments. Today, he is probably best known for inventing roller skates. Some of these contraptions he displayed in the Mechanical Museum, but he kept two prize creations in his workshop in the attic above the museum: two miniature female automata, no more than a foot or two tall. One creature walked across a four-foot space, holding an eyeglass and bowing respectfully toward the onlookers. The other was a dancer holding an animated bird.

Conventional historical accounts are typically oriented around Great Events: battles fought, treaties signed, speeches delivered, elections won, leaders assassinated. Or the textbooks follow the long arc of incremental change: the rise of democracy or industrialization or civil rights. But sometimes history is shaped by chance encounters, far from the corridors of power, moments when an idea takes root in someone's head and lingers there for years until it makes its way onto the main stage of global change. One of those encounters happens in 1801, when a mother brings her precocious eight-year-old son to visit Merlin's museum. His name is Charles Babbage. The old showman senses something promising in the boy and offers to take him up to the attic to spark his curiosity even further. The boy is charmed by the walking lady. "The motions of her limbs were singularly graceful," he would recall many years later. But it is the dancer that seduces him. "This lady attitudinized in a most fascinating manner," he writes. "Her eyes were full of imagination, and irresistible."

The encounter in Merlin's attic stokes an obsession in Babbage, a fascination with mechanical devices that convincingly emulate the subtleties of human behavior. He earns degrees in mathematics and astronomy as a young scholar, but maintains his interest in machines by studying the new factory systems that are sprouting across England's industrial north. Almost thirty years after his visit to Merlin's, he publishes a seminal analysis of industrial technology, *On the*

Economy of Machinery and Manufactures, a work that would go on to play a pivotal role in Marx's *Das Kapital* two decades later. Around the same time, Babbage begins sketching plans for a calculating machine he calls the Difference Engine, an invention that will eventually lead him to the Analytical Engine several years later, now considered to be the first programmable computer ever imagined.

We don't know if the eight-year-old Babbage made a notable impression on Merlin himself. The showman died two years after Babbage's visit, and his collection of wonders—including the captivating automata—were sold to a rival named Thomas Weeks, who ran his own museum a few blocks away on Great Windmill Street. Weeks never put the dancer or the walking lady on display; they remained in his attic, gathering cobwebs until Weeks himself died in 1834, and the entire lot was put up for auction. Somehow, after all those years, Babbage found his way to the auction and purchased the dancer for thirty-five pounds. He refurbished the machine and put it on display in his Marylebone town house, a few feet away from the Difference Engine. In a sense, the two machines belonged to different centuries: the dancer was the epitome of Enlightenment-era fantasy; the Difference Engine an augur of late twentieth-century computation. The dancer was a thing of beauty, an amusement, a folly. The engine was, as its name suggested, a more serious affair: a tool for the age of industrial capitalism and beyond. But according to Babbage's own account, the passion for mechanical thinking that led to the Difference Engine began with that moment of seduction in Merlin's attic, in the "irresistible eyes" of a machine passing for a human for no good reason other than the sheer delight of the illusion itself.

Delight is a word that is rarely invoked as a driver of historical change. History is usually imagined as a battle for survival, for power, for freedom, for wealth. At best, the world of play and amusement belongs to the sidebars of the main narrative: the spoils of progress, the surplus that civilizations enjoy once the campaigns for freedom and affluence have been won. But imagine you are an observer of social and technological trends in the second half of the eighteenth century, and you are trying to predict the truly seismic developments that would define the next three centuries. The programmable pen of Jaquet-Droz's writer—or Merlin's dancer and her "irresistible eyes"—would be as telling a clue about that future as anything happening in Parliament or on the battlefield, foreshadowing the rise of mechanized labor, the digital revolution, robotics, and artificial intelligence.

This book is an extended argument for that kind of clue: a folly, dismissed by many as a mindless amusement, that turns out to be a kind of artifact from the future. This is a history of play, a history of the pastimes that human beings have concocted to amuse themselves as an escape from the daily grind of subsistence. This is a history of what we do for fun. One measure of human progress is how much recreational time many of us now have, and the immensely varied ways we have of enjoying it. A time traveler from five centuries ago would be staggered to see just how much real estate in the modern world is devoted to the wonderlands of parks, coffeeshops, sports arenas, shopping malls, IMAX theaters: environments specifically designed to entertain and delight us. Experiences that were once almost exclusively relegated to society's elites have become commonplace to all but the very poorest members of society. An average middle-class family in Brazil or Indonesia takes it for granted that their free time can be spent listening to music, marveling at elaborate special effects in Hollywood movies, shopping for new fashions in vast palaces of consumption, and savoring the flavors of cuisines from all over the world. Yet we rarely pause to consider how these many luxuries came to be a feature of everyday life.

History is mostly told as a long fight for the necessities, not the luxuries: the fight for freedom, equality, safety, self-governance. Yet the history of delight matters, too, because so many of these seemingly trivial discoveries ended up triggering changes in the realm of Serious History. I have called this phenomenon "the hummingbird effect": the process by which an innovation in one field sets in motion transformations in seemingly unrelated fields. The taste for coffee helped create the modern institutions of journalism; a handful of elegantly decorated fabric shops helped trigger the industrial revolution. When human beings create and share experiences designed to delight or amaze, they often end up transforming society in more dramatic ways than people focused on more utilitarian concerns. We owe a great deal of the modern world to people doggedly trying to solve a high-minded problem: how to construct an internal combustion engine or manufacture vaccines in large quantities. But a surprising amount of modernity has its roots in another kind of activity: people mucking around with magic, toys, games, and other seemingly idle pastimes. Everyone knows the old saying "Necessity is the mother of invention," but if you do a paternity test on many of the modern world's most important ideas or institutions, you will find, invariably, that leisure and play were involved in the conception as well.

Although this account contains its fair share of figures like Charles Babbage—well-to-do Europeans tinkering with new ideas in their parlors—it is not just a story about the affluent West. One of the most intriguing plot twists in the story of leisure and delight is how many of the devices or materials originated outside of Europe: those mesmerizing automata from the House of Wisdom, the intriguing fashions of calico and chintz imported from India, the gravity-defying rubber balls invented by Mesoamericans, the clove and nutmeg first tasted by remote Indonesian islanders. In many ways, the story of play is the story of the emergence of a truly cosmopolitan worldview, a world bound together by the shared experiences of kicking a ball around on a field or sipping a cup of coffee.

The pursuit of pleasure turns out to be one of the very first experiences to stitch together a global fabric of shared culture, with many of the most prominent threads originating outside Western Europe.

I should say at the outset that this history deliberately excludes some of life's most intense pleasures—including sex and romantic love. Sex has been a central force in human history; without sex, there is no human history. But the pleasure of sex is bound up in deep-seated biological drives. The desire for emotional and physical connections with other humans is written into our DNA, however complex and variable our expression of that drive may be. For the human species, sex is a staple, not a luxury. This history is an account of less utilitarian pleasures; habits and customs and environments that came into being for no apparent reason other than the fact that they seemed amusing or surprising. (In a sense, it is a history that follows Brian Eno's definition of culture as "all the things we don't *have* to do.") Looking at history through this lens demands a different emphasis on the past: exploring the history of shopping as a recreational pursuit instead of the history of commerce writ large; following the global path of the spice trade instead of the broader history of agriculture and food production. There are a thousand books written about the history of innovations that came out of our survival instincts. This is a book about a different kind of innovation: the new ideas and technologies and social spaces that emerged once some of us escaped from the compulsory labor of subsistence.

The centrality of play and delight does not mean that these stories are free of tragedy and human suffering. Some of the most appalling epochs of slavery and colonization began with a new taste or fabric developing a market, and unleashed a chain of brutal exploitation to satisfy that market's demands. The quest for delight transformed the world, but it did not always transform it for the better.

In 1772, Samuel Johnson paid a visit to one of the predecessors of Merlin's Mechanical Museum, a showcase run by an engineer named James Cox, who became one of Merlin's mentors. Exploring Cox's exhibition was like walking through the pages of al-Jazari's illustrated book: the rooms were filled with animated elephants, peacocks, and swans, glittering with jewels. Johnson published an account of his visit in the *Rambler.* "It may sometimes happen," he wrote, "that the greatest efforts of ingenuity have been exerted in trifles; yet the same principles and expedients may be applied to more valuable purposes, and the movements, which put into action machines of no use but to raise the wonder of ignorance, may be employed to drain fens, or manufacture metals, to assist the architect, or preserve the sailor."

In other words, the ingenious "trifles" of the automata often serve as a kind of augur of more substantial developments to come. This foreshadowing effect is clearly visible in the commentary that built up around the great automata of the eighteenth century: Jaquet-Droz's Writer, Vaucanson's duck, the famous chess-playing "Mechanical Turk" originally designed in the 1770s by the Hungarian inventor Wolfgang von Kempelen. (The Turk turned out to be less of a mechanical achievement, as the chess was actually played by a man hidden inside the contraption.) While these contraptions sparked amazement and debate in their prime—several essays were published in the late 1700s trying to solve the mystery of the Turk's chess abilities—they reached their cultural peak in the middle of the nineteenth century, well after most of their showcases had gone out of business. The automata inspired Marx's theories on the future of labor and propelled Babbage toward his prophetic vision of mechanized intelligence. They planted the seed for Mary Shelley's *Frankenstein.* Edgar Allan Poe's attempt to explain the secrets of the Mechanical Turk laid the groundwork for his invention of the detective story. The automata were animated by the scientific and engineering knowledge of the eighteenth century, but they unleashed broader hopes and fears that belonged properly to the nineteenth. In both their mechanical design and their philosophical implications, the automata were ahead of their time.

This phenomenon turns out to appear consistently throughout the history of humanity's trifles. The guilty pleasures of life often give us a hint of future changes in society, whether those pleasures take the form of English ladies shopping for calico fabrics in London in the late 1600s, or ancient Roman feasts laden with spices from the far corners of the globe, or carnival hucksters promoting strange optical devices that create the illusion of moving pictures, or computer programmers at MIT in the 1960s playing Spacewar! on their million-dollar mainframes. Because play is often about breaking rules and experimenting with new conventions, it turns out to be the seedbed for many innovations that ultimately develop into much sturdier and more significant forms. The institutions of society that so dominate traditional history—political bodies, corporations, religions—can tell you quite a bit about the current state of the social order. But if you are trying to figure out what's coming next, you are often better off exploring the margins of play: the hobbies and curiosity pieces and subcultures of human beings devising new ways to have fun. "Each epoch dreams the one to follow, creates it in dreaming," the French historian Michelet wrote in 1839. More often than not, those dreams do not unfold within the grown-up world of work or war or governance. Instead, they emerge from a different kind of space: a space of wonder and delight where the normal rules have been suspended, where people are free to explore the spontaneous, unpredictable, and immensely creative work of play. You will find the future wherever people are having the most fun.

1

Fashion and Shopping

The Calico Madams

The sea snail *Hexaplex trunculus* lives in shallow waters and tidal pools along the coast of the Mediterranean, and along the shores of the Atlantic, from Portugal down to the western Sahara. To the untrained eye, the murex snail, as it is also called, looks like an ordinary mollusk, housed in a conical shell ringed by bands of spikes. Millions of years ago, the snail evolved a kind of bioweapon used to sedate prey and defend itself against predators: an inky secretion that contains a rare compound called dibromoindigo. Almost four thousand years ago, the Minoan civilization based in the Aegean islands discovered that the murex snail secretion could be used as a dye to create one of the rarest of shades: the color purple.

Over time, the purple dye took on the name of a town in southern Phoenicia, Tyre, where it was mass-produced. The exact procedure for manufacturing Tyrian purple is unknown today, although Pliny the Elder included a fragmentary recipe in his *Natural History*. Modern attempts to re-create the dye suggest that more than ten thousand snails were required to produce just one gram of Tyrian dye. But if the production techniques remain a mystery, the historical record is clear about one thing: Tyrian purple endured as a symbol of status and affluence for at least a thousand years. Bands of Tyrian purple were woven into the tunics of Roman senators; a child conceived by one of the emperors of Byzantium was given the honorific *Porphyrogenitus*—literally, "born in the purple." Over the millennium that passed from the age of the Phoenicians to the fall of Rome, an ounce of Tyrian purple dye was worth significantly more than an ounce of gold, a valuation that compelled sailors to explore the entire coastline of the Mediterranean for colonies of murex snails.

Eventually, though, the supply of *Hexaplex trunculus* in the Mediterranean could not keep up with the demand for Tyrian purple, and a few intrepid Phoenician sailors began to contemplate more ambitious voyages in search of the mollusk, beyond the placid waters of the inland sea, out onto the gray, turbulent waves of the Atlantic itself. The Phoenicians had already passed through the Strait of Gibraltar in search of alluvial tin deposits, their distinctive cedar-planked ships, powered by thirteen oarsmen on each side, hugging the coast of Spain through waters that, technically speaking, belonged to the Atlantic. But it was the murex snail that compelled them to take on the towering waves and uncharted waters of the open ocean. They ventured down the coast of North Africa, where they eventually discovered a bounty of sea snails that would keep the aristocracy cloaked in purple well into the Dark Ages. The legacy of these voyages extends far beyond simple fashion. The passage out of the Mediterranean into the vast mystery of the Atlantic marked a true threshold moment in the history of human exploration. "The Phoenicians' now-proven aptitude for sailing the North African coast was to be the key that unlocked the Atlantic for all time," Simon Winchester writes. "The fear of the great unknown waters beyond the Pillars of Hercules swiftly dissipated." Think of all the ways the world would be transformed by vessels launched from Mediterranean countries, exploring the Atlantic and beyond. Those vessels would eventually leave in search of gold, or religious freedom, or military conquest.

The murex snail

But the first siren song that lured them onto the open ocean was a simple color.

Garment design has driven technological innovation from the very beginning of human existence. Shears, sewing needles, and scrapers for converting animal skins into protective coverings for the body are among the oldest tools recovered from the Paleolithic age. To be sure, much of that innovation was utilitarian in nature. Ascots and hoop skirts aside, most clothing has some functional value, and certainly our ancestors fifty thousand years ago were making clothes with the explicit aim of keeping warm and dry and protected from potential threats. The fact that so much technological innovation—from the first knitting needles to hand looms to the spinning jenny—has emerged out of textile production can seem, at first glance, more a matter of necessity's invention. And yet the archeological record is replete with early examples of purely decorative toolmaking: a shell necklace discovered in the Sikul Cave in Israel was crafted more than a hundred thousand years ago. As soon as humans became toolmakers, they were making jewelry.

Whatever mix of playfulness and practicality drove early human garment design, the invention of Tyrian purple announced a fundamental shift toward delight and surprise—a shift, in a sense, from function to fashion. No one *needs* the color purple. It does not protect you against malaria, or supply useful proteins, or reduce the chances that you will die in childbirth. It just looks nice, particularly if you live in a world where purple garments happen to be rare.

You might reasonably object at this point that those Phoenician snail wranglers—and the oarsmen that first took them past the Strait of Gibraltar—were motivated by financial gain, and not some sublime aesthetic response to purple itself. That is certainly the canonical way of telling it. As soon as we invented liquid currencies, human beings were suddenly willing to take on improbably ambitious and dangerous schemes if the price happened to be right. People left the safety of the Mediterranean because there was money in it—a motivation that was certainly powerful, but not particularly newsworthy.

A comparable argument might be made for the importance of status in the display of those purple garments. Humans evolved in hierarchical societies and most of us acknowledge that status-seeking is a common, if sometimes regrettable, driver of human behavior. The Phoenician aristocracy wanted to dye their clothes in shades of Tyrian purple so they could display their superiority over the commoners, and they were willing to pay for the privilege. Again, the causal chain is a familiar one: people will go to great lengths to satisfy the needs of the ruling elite if they are amply compensated for their labor. The fact that this labor involved harvesting thousands of snails may be an intriguing historical yarn, but does it really tell us anything new about the deep-seated forces that drive historical change?

With all due respect to Occam's razor, I think in this case the simpler story is not correct, or at least it fails to include the most interesting part of the explanation. The financial gain or status symbols were secondary effects; the initial fixation with purple was the prime mover. Take away the purely aesthetic response to the Tyrian dye, and the whole chain of exploration, invention, and profit falls apart. This turns out to be a recurring pattern in the history of play. Because delightful things are valuable, they often attract commercial speculation, which funds and cultivates new technologies or markets or geographic exploration. When we look back at that process, we tend to talk about it in terms of the money and markets or the vanity of the ruling elite driving the new ideas. But the money has its own masters, and in many cases the dominant one is the human appetite for surprise and novelty and beauty. If you dig past the archeological layers of technological invention, profit motive, conquest, and status-seeking, you will often find an unlikely stratum that lies beneath the more familiar layers: the simple pleasure of a new experience—in this case, the red and blue cones of our retinas registering a strange hybrid shade almost never found in nature. Somehow the story gets cast in the retelling as a tale of heroic inventors or efficient capital markets or brutal exploitation. That initial moment of delight becomes an afterthought, a footnote to the master narrative.

Nowhere is this oversight more glaring than in the story behind the greatest technological upheaval of modern times: the industrial revolution.

In the last few decades of the seventeenth century, a new pattern became visible—arguably for the first time in history—on the streets of a few select neighborhoods in London: St. James, Ludgate Hill, Bank Junction. A row of shops, each offering tantalizing collections of fabric, or jewelry, or home furnishings, clustered together on a few city blocks. The shopfronts featured large glass window displays, with merchandise arranged in visually arresting styles. The interiors were festooned with pillars, elaborate mirrors and lighting, sculpted cornices, and draperies. An observer from the early 1700s described them as "perfectly gilded theaters."

All of this theatrical elegance was designed to create a new kind of aura around the simple act of buying goods. Earlier in the seventeenth century, shopping galleries like the New Exchange and Westminster Hall had created bustling, immersive spaces for commercial transactions, but the new shops added a measure of grandeur and elegance that made the galleries seem cramped and oppressive by comparison. In the exchanges, each vendor's space was small and largely unfurnished, closer to the stalls of a traveling fair or street peddlers. The new shops created a much more sumptuous environment, as though the consumer were entering the drawing room of a minor lord instead of bargaining with a street hawker. For the first time, the design of the shop became a part of the marketing message. Indeed, in an age that predates the modern craft of advertising, those shop designs were among the very first forms of marketing ever concocted. "The seductive design of shops was intended to encourage customers to stay and to look around, to see shopping as a leisurely pursuit and an exciting experience," writes historian Claire Walsh. "The more time a customer spent in the shop, the more attentively and persuasively they could be served, and in this sense the design of the shop was very much a part of the sales process." They made the act of shopping an end, and not just a means.

Some contemporary observers, mostly men, denounced the new shops as palaces of deception, designed to weave a spell over their customers. Describing the new fashionable shops in the resort town of Bath in the early 1700s, Abbé Prévost complained that they took

Trade card advertising a London shop, circa 1758

advantage of "a kind of enchantment which blinds everyone in these realms of enjoyment, to sell for their weight in gold trifles one is ashamed of having bought after leaving the place." In his 1727 survey of British commercial practice, *The Compleat English Tradesman*, Daniel Defoe devoted an entire chapter to the new practice of outfitting shops with such lavish trappings, a custom that appears to have baffled Defoe: "It is a modern custom, and wholly unknown to our ancestors . . . to have tradesmen lay out two-thirds of their fortune in fitting up their shops . . . in painting and gilding, fine shelves, shutters, boxes, glass-doors, sashes, and the like," he wrote. "The first inference to be drawn from this must necessarily be, that this age must have more fools than the last: for certainly fools only are most taken with shows and outsides."

Defoe ultimately decided that there must be some kind of functional motive behind these seemingly excessive displays: "Painting and adorning a shop seems to intimate, that the tradesman has a large stock to begin with; or else the world suggests he would not make such a show." Defoe's perplexity here is almost touching: you can see his mind working in overdrive to come up with a logical explanation for the frivolities of fine shelves and sashes. From our modern perspective, we can see clearly how the messaging embodied in the lavish shop decor obviously signaled more than just a large inventory; it created an envelope of luxury and high fashion that elevated the act of shopping itself into a form of entertainment. The consumers flocking to these new commercial spaces weren't just there for the goods they could purchase. They were there for the wonderland of the space itself.

Where shopping for clothes had previously been a straightforward, no-frills series of exchanges, bartering with street vendors or tradesmen—no different from buying eggs or milk—now the practice of browsing and "window shopping" became its own sought-after experience. Before the rise of these lavish London shops, one went to market when one had something specific to purchase. Bazaars and open-air markets existed, of course, but they lacked the sumptuous displays of these new London shops. These "perfectly gilded theaters" transformed the journey of shopping into its own reward. A 1709 contributor to *Female Tatler* describes the phenomenon—now ubiquitous in the developed world—with fresh eyes: "This afternoon some ladies, having an opinion of my fancy in cloaths, desired me to accompany them to Ludgate-hill, which I take to be as agreeable an amusement as a lady can pass away three or four hours in."

The language here—"agreeable amusements"—doesn't fully do justice to the eventual magnitude of the transformation it was describing. It was a subtle shift, hard to notice if you weren't a proprietor of one of these shops, or a customer. To the untutored eye, those shops seemed like just a minor twist on the peddlers and tradesmen who had sold goods in the city for hundreds of years. In fact, the shift was so subtle that very few records were kept to document its existence. Like so many cultural revolutions that would follow, the modern experience of shopping trickled into the world as a minor subculture, enjoyed by a tiny fraction of the overall population, ignored by the mainstream—until one day when the mainstream woke and found that it had been profoundly redirected by this strange new tributary. Every now and then, the creek floods the river.

The lack of historical records meant that, until recently, most cultural historians assumed that the birth of consumer culture and the sensuous excesses of shop displays began in the late nineteenth century with the invention of the department store, after the first wave of industrialization. But in fact, that traditional story has it exactly backward: the trivial pursuits of shopping were not a secondary effect of the industrial revolution and the rise of bourgeois consumer culture. In several key respects, those elaborate drapers' shops on the streets of London helped *create* the industrial revolution. And that is because those gilded theaters were increasingly designed to showcase the brilliantly colored calico prints of a miraculous new material from the other side of the world: cotton.

Archeologists believe that the practice of domesticating the plant *Gossypium malvaceae*, and weaving it into the fabric we now call cotton, dates back more than five thousand years. Interestingly, the manufacture of cotton, using primitive combs and hand spindles, appears to have been independently discovered in four different places around the world, roughly at the same time: in the Indus Valley of present-day Pakistan, in Ethopia, along the Pacific coast of South America, and somewhere in Central America. The utility of *Gossypium malvaceae*'s fibers seems to become apparent to any sufficiently advanced civilization situated in an ecosystem where the plant naturally lives. Some of those early civilizations failed to invent writing or wheeled vehicles, but they did figure out a way to turn the thin fibers of the cotton boll into soft and breathable fabrics.

Until the 1600s, those fabrics were largely mythical to most Northern Europeans, who dressed themselves in thicker, scratchy garments of wool and linen. Cotton was so fanciful that the globe-trotting British knight John Mandeville famously described in the 1300s "a wonderful tree [in India] which bore tiny lambs on the endes of its branches. These branches were so pliable that they bent down to allow the lambs to feed when they are hungrie."

But the soft texture of cotton would prove to be only part of its appeal. After thousands of years of experimentation, Indian dyers located on the Coromandel coast established an elaborate system of soaking vibrant dyes like madder and indigo into the fabric, employing lemon juice, goat urine, camel excrement, and metallic salts.

Workers printing and painting calico

Most colored fabrics in Europe would lose their pigment after a few washings, but the Indian fabrics—called chintz and calico—retained their color indefinitely. When Vasco da Gama brought back a cargo full of textiles in 1498 from his landmark expedition around the Cape of Good Hope, he gave Europeans their first real experience of the vivid patterns and almost sensual textures of calico and chintz.

As fabrics, calico and chintz first made their way into the routine habits of Europeans through the gateway drug of interior decorating. Starting in the 1600s, the well-to-do of London and a few other European cities began festooning their drawing rooms and *boudoirs* in the floral or geometric patterns of calico cloth. As clothing itself, cotton was initially perceived to be too light for the climate of Northern Europe, particularly in the winter. But in the final decades of that century, a strange feedback loop began to resonate among the fashionable elite of London society. They began to crave cotton on their bodies. Drapes were cut down and converted into dresses, settees plundered to sew into jackets or blouses. Perhaps most important, cotton undergarments that could be worn in the depths of winter, buffering the skin from the irritations of wool, became an essential element of a lady's wardrobe.

The surge in interest in Indian textiles was a tremendous boon for the East India Company, which went from importing a quarter of a million pieces in 1664 to 1.76 million twenty years later. (More than 80 percent of the company's trade was devoted to calico at the height of the craze.) But the news was not as encouraging for England's native sheep farmers and wool manufacturers, who suddenly saw their livelihoods threatened by an imported fabric. The craze for cotton was so severe that by the first decade of the next century it triggered a kind of moral panic among the rising commentariat, accompanied by a series of parliamentary interventions. Hundreds if not thousands of pamphlets and essays were published, many of them denouncing the "Calico Madams" whose scandalous taste for cotton was undermining the British economy. "The Wearing of printed Callicoes and Linnens, is an Evil with respect to the Body Politick," one commenter announced. Defoe himself wrote multiple screeds on what he considered "a Disease in Trade . . . a Contagion, that if not stopp'd in the Beginning, will, like the Plague in Capital City, spread itself o'er the whole Nation." Plays, poems, and popular songs were composed decrying the spread of calico. One song, "The Spittle-Fields Ballad" (which took its name from a neighborhood heavily populated by weavers) took the public shaming to an extreme: "none shall be thought / A more scandalous Slut / Than a taudry Callico Madam." Rioting weavers marched on Parliament and ransacked the home of the East India Company's deputy governor. One can safely assume that at no other point in human history have women's undergarments provoked so much patriotic fury.

Responding to the outrage, Parliament passed a number of protectionist acts, starting with a ban on imported dyed calicos in the 1700s, which left open a large loophole for traders to import raw cotton fabrics, to be dyed on British shores. In 1720, Parliament took the more draconian step of banning calico outright, via "An Act to Preserve and Encourage the Woollen and Silk Manufactures of this Kingdom, and for more Effectual Employing the Poor, by Prohibiting the Use and Wear of all Printed, Painted, Stained or Dyed Callicoes in Apparel, Household Stuff, Furniture, or otherwise."

Ironically, the fears that ladies' fashion trends would undermine the British economy turned out to have it exactly backward. The immense value of the cotton trade had already set a generation of British inventors off in search of mechanical tools that could mass-produce cotton fabrics: beginning with John Kay's flying shuttle, patented in 1733, followed several decades later by Richard Arkwright's spinning (or water) frame, then Eli Whitney's cotton gin, not to mention the endless refinements to the steam engine rolled out during the 1700s, many of which were originally designed to enhance textile production. (Steam engines would eventually power a wide range of industrial production and transportation, but their initial application was dominated by mining and textiles.) Instead of deflating the British economy, the Calico Madams unleashed an age of British industrial and economic might that would last for more than a century.

That cotton changed the world is indisputable. The more interesting question is how this intense appetite for cotton came about. The traditional explanation held that cotton conquered Europe thanks to its intrinsic virtues as a textile and to its price. Yet the historian John Styles has demonstrated that cotton failed to penetrate a true mass market until well into the nineteenth century, and was generally more expensive than the rival products of wool and

W G Taylor's Power Loom and Patent Calico Machine

linen. What set cotton apart was not practical matters of cost and comfort but rather the more ethereal trends of fashion. "The spectacular early triumph of cotton depended most of all on its visual, decorative, fashionable qualities," Styles writes. "Where appearance was crucial, cotton succeeded. Where utilitarian durability counted, cotton sometimes lagged behind."

A few perceptive individuals at the time were able to see beyond the Calico Madam protectionist outrage and detect the deeper trends lying beneath the craze for cotton. The economist and financial speculator Nicholas Barbon observed in his 1690 work, *A Discourse of Trade*: "It is not Necessity that causeth the Consumption. Nature may be Satisfied with little; but it is the wants of the Mind, Fashion and the desire of Novelties and Things Scarce that causeth Trade." But how did that fashion and that "desire of Novelties" spread through European society? Recall that da Gama first brought calico in bulk to Europe in 1498. Yet almost two centuries passed before a critical mass of people began draping their bodies with it. What caused cotton's eventual takeoff? This was, of course, an age where advertising and image-based media were literally nonexistent. There were no *Vanity Fair* spreads or Fashion Week broadcasts to get the word out. You could only experience calico through direct encounters: on the bodies or furniture of people you knew. For a century and a half, that's how the taste for cotton spread through the population, one banquet at a time. But fads big enough to transform global economies don't tend to self-organize out of casual gossip. They usually require some kind of amplifier.

Starting in the second half of the seventeenth century, that amplifier appeared on the streets of Ludgate Hill and St. James: those luxurious shopfronts, attracting the eyes of women with enough wealth to spend a few hours browsing for goods that they didn't, technically, need. Calico had been circulating through Northern Europe for a hundred and fifty years, but it didn't turn into a true frenzy until the new rituals of shopping—the window displays, the clustered stores, the lavish interiors—had come into being. It's possible that the news of cotton's charms simply passed through word-of-mouth networks and slowly built up in intensity. But the historical congruence of those high-end London stores and the calico frenzy of the late 1600s strongly suggests that the Calico Madam was herself the by-product of a new kind of marketplace and the new recreational pastime of shopping. The shopkeepers made the cotton revolution just as much as da Gama did.

The distinction matters because of that standard theory about the rise of "consumer society" and its relationship to industrialization. When historians have gone back to wrestle with the question of *why* the industrial revolution happened, when they have tried to define the forces that made it possible, their eyes have been drawn to more familiar culprits on the supply side: technological innovations that increased industrial productivity, the expansion of credit networks and financing structures; insurance markets that took significant risk out of global shipping channels. But the frivolities of shopping have long been considered a secondary effect of the industrial revolution itself, an effect, not a cause; a cultural appetite that wouldn't be whetted until the rise of nineteenth-century department stores like Le Bon Marché and Macy's. According to the standard theory, industrialization created mechanical processes that greatly reduced the cost of manufacturing and transporting goods, and built up a base of upper-middle-class citizens with enough spare cash to drop on the niceties—which then led to the birth of consumerism. But the Calico Madams suggest that the standard theory is, at the very least, more complicated than that: the "agreeable amusements" of shopping most likely came first, and set the thunderous chain of industrialization into motion with their seemingly trivial pursuits.

This might seem like an academic distinction, but the stakes in this particular wager happen to be high ones. At its core is the question of why big changes in society happen. Are they driven exclusively by new tools and cultural practices that satisfy existential needs, like nutrition, shelter, or sexual reproduction? Or are they also driven by more mercurial appetites? And even if you limit the frame of reference to the Industrial Revolution itself, the story of those luxury stores and the delightful patterns of calico cloth has real weight to it. It strongly suggests that the conventional narrative of industrialization is flawed both in terms of the sequence of events *and* the key participants. The great takeoff of industrialization, for instance, has inevitably been told as the work of European and North American men—heroes and villains both—building steam engines and factories and shipping networks. But those dyers tinkering with calico prints on the Coromandel coast, creating new designs for the sheer beauty of it; those English women enjoying the "agreeable amusements" of shopping on Ludgate Hill—these were all active shapers of the modern reality of industrialization, as important, in a way, as the James Watts and Eli Whitneys of conventional history.

The account is necessarily murky because so few contemporaries found it necessary to take note of these new shopfronts until the calico craze had threatened to decimate the English economy. And in a way, those omissions were understandable. This was the age of Oliver Cromwell and the Glorious Revolution; larger, more masculine struggles, featuring the traditional agents of world history—kings, armies, priests—were surging across Europe and the British Isles. But with perfect hindsight, if you were sitting there in 1680 trying to predict the massive changes coming to global capitalism, you couldn't have found a better crystal ball than those calico shops in London.

You can still hear the echoes of cotton's big bang, even in the distant galaxy of contemporary American politics.

During the Cretaceous period, roughly a hundred million years ago, the body of water that eventually drained down to the become the modern Gulf of Mexico covered the southern half of Georgia and Alabama, forming a crescent coastline. The varied marine life that lived in those waters left behind a rich, dark soil. Millions of years after those seas had vanished from the landscape, American farmers discovered those soils were particularly well suited for growing cotton, and a long, curving arc of plantations took root on the scene of that long-forgotten coastline. Starting in the early 1800s, plantation owners forcibly migrated millions of slaves to work what became known as the black belt, a term derived from both the color of the slaves' skin and the color of the soil itself.

The legacy of that forced migration still lingers: in the 2008 U.S. presidential election, Barack Obama's support in the south followed almost exactly the distinct crescent of the black belt, thanks to the large number of African-Americans still residing in those counties two centuries later. The ultimate explanation of *why* Obama won those counties forces us to look beyond the present-tense politics and tell a longer story: from the ancient geological forces that deposited that crescent of black soil, to the appetite for cotton stirred up by the shopkeepers of London, to the brutal exploitation of the plantation system engineered to satisfy that new demand.

The story of calico and chintz is a chilling reminder that the amusements of life have often triggered some of the worst atrocities in history. The sensual delight of these new fabrics inspired a wave of entrepreneurial activity and technological ingenuity, but it also unleashed some of the most destructive forces that the world had ever seen. Cotton dug scars across the face of the earth that are still healing, three hundred years later. The most visible, and painful, of these was slavery in the American south.

Slaves had been a part of the American society from the first decades of the colonial era, but it was cotton that turned the forced labor of African-Americans into the cornerstone of the south's economy. In 1790, only a handful of counties along the seaboard of the American southeast had a slave population higher than 50 percent, mostly working tobacco plantations. By 1860, slaves made up roughly half of the entire population of the southern states, numbering more than five million in total. The cost of cotton can be measured in military deaths as well. Without its economic dependence on cotton plantations, the south might well have grudgingly acceded to the abolitionist movement, and the Civil War—a war that took as many American lives as all other U.S. military conflicts combined—might have never happened.

Slavery was cotton's most appalling legacy, but in the new factories that sprouted first in England and then in the northern states of the United States, the lives of the workers turning the raw cotton into textiles was also horrific: the first years of the industrial age saw decreasing life spans, child-labor abuse, seventy-hour workweeks, air

Fashion plate from The Lady's Magazine, *1774*

pollution that resembled modern-day Beijing, and the dismantling of many of the conventions of agrarian life.

Between slavery and the grotesque working conditions of early industrialization, one can make the argument that the desire for cotton was the single worst thing to happen to the planet between 1700 and 1900—that it provoked more suffering than any other new development in that period. Those early shoppers savoring the soft fabrics and vibrant patterns of calico at the end of the seventeenth century would have been baffled to know that their fashion decisions were going to unleash such traumatic and deadly forces across the globe.

By the end of the eighteenth century, the calico craze—supercharged by the mechanical looms of industrial England—had given birth to the first true fashion "industry." A woman in 1770s London could enjoy dozens of illustrated periodicals devoted to the latest fabrics or gowns or hairstyles. These publications lacked the snazzier titles of the modern age; instead of *Elle* or *Cosmopolitan*, the titles included the *Ladies' Pocket Book*; the *General Companion to the Ladies, or Useful Memorandum Book*; the *Ladies' New and Elegant Pocket Book*; the *Ladies' Pocket Journal*; and the *Polite and Fashionable Ladies' Companion*. (The first full-color fashion image appeared in a publication called *The Lady's Magazine* in 1771.) Those magazine titles hint at the complex gender politics of the fashion revolution. On the one hand, women were in many ways the prime movers behind the vast commercial upheaval of the age as advertisers, publishers, textile manufacturers, and financial speculators all set off in pursuit of this strange, seemingly arbitrary need for the latest vogue. On the other hand, the proliferation of guidebooks helped cement an image of female identity that seems, at face value, to be undeniably regressive: that being a woman in modern society entailed tracking the increasingly complex and ever-shifting winds of taste and convention.

Propelled by these publications and by the commercial interests of the manufacturers, the metronome of fashion sped up markedly in the second half of the eighteenth century. As late as 1723, Bernard Mandeville could observe, contemptuously, that most fashion "modes seldom last above Ten or Twelve Years"—a decade-long cycle being a sign of shameful transience. Starting in the 1750s, the pace accelerated to a semiannual rotation, and by the 1770s, fashion more or less stabilized at its current rate of change, where each year introduces a new "in" look. The historians Neil McKendrick, John Brewer, and J. H. Plumb document the change in their groundbreaking work *The Birth of A Consumer Society*: "In 1753 purple was the in-colour," McKendrick writes. "In 1757 the fashion was for white linen with a pink pattern. In the 1770s the changes were run even more rapidly—in 1776 the fashionable color was 'couleur de Noisette,' in 1777 dove grey, in 1779 'the fashionable dress was haycock stain trimmed with fur.' By 1781 'stripes in silk or very fine cambric-muslin' were in." For the first time, the cities of Europe—followed, eventually, by the rest of the world—began to be synchronized to a kind of aesthetic clock, with regular pulses of new instructions rippling out from the metropolitan centers: now white linen, now dove gray, now haycock stain. Alongside the ancient rhythms of the seasons, and the more recent emergence of boom-and-bust economic cycles, a new rhythm appeared: the choreographed oscillations of *la mode*.

Was this, in the end, a good thing? Certainly one can make the argument that fashion compelled a great number of people to spend their days trying to ascertain if their new gown was in fact the correct shade of "couleur de Noisette," when they might have been attending to more important things. Over time, the social enchantment of fashionable clothes came to envelop almost the entire object world as the advertising trade took the techniques it had developed to sell style and opulence and directed it to more mundane products. The arbitrary linking of status and affluence with shades of fabric—which began in part with Tyrian purple—metastasized into a kind of permanent fantasyland where everything for sale was packaged with a supplementary aura. The British cultural historian Raymond Williams saw this as a form of magical thinking:

> *It is often said that our society is too materialist, and that advertising reflects this . . . But it seems to me that in this respect our society is quite evidently not materialist enough, and that this, paradoxically, is the result of a failure in social meaning, values and ideals . . . If we were sensibly materialist, in that part of our living in which we use things, we should find most advertising to be of an insane irrelevance. Beer would be enough for us, without the additional promise that in drinking it we show ourselves to be manly, young in heart, or neighborly. A washing-machine would be a useful machine to wash clothes, rather than an indication that we are forward-looking or an object of envy to our neighbors.*

But the legacy of the fashion revolution is not just the illusory emotional attachments that have now come to envelop everything we buy, from dresses to phones to pickles. In an important sense, those absurd-sounding garments were a democratizing force. Until the fashion revolution of the eighteenth century, a wide gulf separated the style of the rich and the rest of society. Elaborate fashions, like Tyrian purple, functioned as a kind of decorative shibboleth separating the elites from the common folk. But the manufacturing and advertising businesses that brought those fashions to upper-middle-class ladies in the late 1700s were breaking down an important wall. You could now "pass" for an aristocrat on the street or in the parlor if you paid attention to the signs and symbols circulating through the *Lady's Pocket Book* or the *Polite and Fashionable Ladies*

Companion. Today, we don't blink an eye when some of the richest people in the world dress like your average adolescent skate punk, in hoodies and jeans and old Adidas sneakers. But in the late 1700s, deliberately blurring the semiotics of class was a novel phenomenon, one that seemed to confound expectations about different social stations.

Did blurring the sartorial lines between aristocracy and middle class help usher in the genuine political reform that led to the more egalitarian societies of the nineteenth and twentieth centuries? Certainly many conservative observers at the time were convinced that these fashion changes were dismantling the proper divisions between the classes. In 1763, the *British Magazine* observed the democratization of dress that was under way, and grumbled: "The present rage of imitating the manners of high life hath spread itself so far among the gentle folks of lower life, that in a few years we shall probably have no common folks at all." Another griped in 1756, "It is the curse of this nation that the laborer and the mechanic will ape the lord; the different ranks of people are too much confounded: the lower orders press so hard on the heels of the higher, if some remedy is not used the Lord will be in danger of becoming the valet of his Gentleman." The *London Magazine* warned in 1772 that the "lower orders of the people (if there are any, for the dissections are now confounded) are equally immersed in their fashionable vices."

In his epic account of the rise of global capitalism, *Civilization and Capitalism*, Fernand Braudel suggested that the appetite for fashion—despite its apparent superficiality—had unsuspected depth. "Is fashion in fact such a trifling thing?" he asked, in a much-quoted passage. "Or is it, as I prefer to think, rather an indication of deeper phenomena—of the energies, possibilities, demands and joie de vivre of a given society, economy, and civilization? . . . Can it have been mere coincidence that the future was to belong to the societies fickle enough to care about changing the colors, materials and shapes of

Aristide Boucicaut

costume, as well as the social order and the map of the world—societies, that is, which were ready to break with their traditions." For Braudel, fashion was both a symptom and cause of a certain social restlessness, a willingness to challenge existing conventions. Sometimes those conventions took the form of political or social hierarchies, laws, or voting rights. Sometimes they took the form of a haycock stain trimmed with fur.

If the eighteenth century encouraged an increasingly democratic model of fashion, the nineteenth century witnessed the creation of a kind of people's palace in the form of the department store, the immense new spaces that conjured up, for the first time, an entire phantasmagoria of consumption. The origins of the department store are somewhat contested by historians of consumer society: some point to Bainbridge's in London, others to Alexander Turner Stewart's "Marble Palace" in downtown Manhattan. But no store from the period captured the imagination of the world like Le Bon Marché, founded by a French fabric salesman named Aristide Boucicaut.

Boucicaut had run a modest shop originally named Au Bon Marché—which means both a "good market" and a "good deal" in French—since 1838, but didn't begin to dream of a more elaborate enterprise until a visit to the World's Fair of 1855 suggested to him the seductive power of sensory overload. Boucicaut's creation would not just be a store outfitted with the trappings of style and luxury, as in those original seventeenth-century shops; it would be an entire wonderland of clothes, fabrics, furniture, trinkets, jewelry, and countless other goods—all collected together in one extravagant space.

Before Le Bon Marché, shopkeepers had competed for the customer's dollar either by creating organized, efficient spaces where the goods for sale were presented in a coherent display that allowed the customer to acquire what they needed and move on, or they presented a fashionable, luxurious mise-en-scène that gave shopping the allure of an upper-class lifestyle. But Boucicaut was the first to recognize the commercial potential of disorientation and overload. "What's necessary," he explained of his customers, "is that they walk around for hours, that they get lost. First of all, they will seem more numerous. Secondly . . . the store will seem larger to them. And lastly, it would really be too much if, as they wander around in this organized disorder, lost, driven, crazy, they don't set foot in some departments where they had no intention of going, and if they don't succumb at the sight of things which grab them on the way."

In this calculated confusion, Boucicaut was adapting one of the classic archetypes of the nineteenth-century city—the flaneur, Baudelaire's "man of the crowd"—to the emerging customs of mass consumption. The sensory shocks and stimuli of dense urban centers had created a new pastime among aesthetes during this era: the aimless stroll through surging crowds, chattering cafés, and bustling markets of the big city, an experience that Baudelaire likened to a "kaleidoscope gifted with consciousness." The art of flânerie was, in its way, a kind of urban version of the Romantic sublime; instead of experiencing sensory overload contemplating the majestic peaks of the Matterhorn, one achieved a similar sensibility immersing oneself in the ocean of humanity flowing down the rue du Bac.

Boucicaut recognized that this undirected delight could be channeled to commercial ends, if one created an environment with sufficient sensory excess. And so he set about building a shopping space that would be unlike anything that had come before it, a "cathedral of commerce," in his own self-congratulatory words. "Dazzling and sensuous, the Bon Marché became a permanent fair, an institution, a fantasy world, a spectacle of extraordinary proportions, so that going to the store became an event and an adventure," the historian Michael Miller writes. "One came now less to purchase a particular article than simply to visit, buying in the process because it was part of the excitement, part of an experience that added another dimension to life." Boucicaut's ambition was so vast that he required new engineering techniques to bring his vision to life. Hiring Gustave Eiffel as his main engineer, Boucicaut built his store around a cast-iron framework topped by enormous skylights, creating oversized galleries illuminated with natural daylight. (The modern shopping experience owes a great debt to materials science and engineering: not just in the form of those cast-iron frames, but also new techniques that allowed giant plate-glass displays to tempt passersby on the sidewalks.) As always, the innovations were all in the service of facilitating consumption: the grand salons enabled large crowds to flow through the store more easily, and gave Boucicaut a massive tableau to stage his extravagant spectacles. Interior balconies allowed visitors to gaze down at the throngs surging through the main floor. By the time of its completion in 1887, the store took up 52,800 square meters, leading many commentators, Boucicaut included, to dub it the eighth wonder of the world. The entrance on the rue de Sèvres was majestically positioned under a cupola, and festooned with caryatids and statues of Roman gods. "The impression was that of entering a theater," Miller writes, "or perhaps even a temple."

Temples and cathedrals—religious metaphors abound in descriptions of Le Bon Marché from the period. Perhaps the most famous observer of the department store's dazzling and disorienting impact was Émile Zola, whose 1883 novel *Au Bonheur des Dames* tells the story of a Bon Marché equivalent, created by a thinly disguised stand-in for Boucicaut named Octave Mouret. In one famous passage, Zola suggests a kind of zero-sum relationship between the department stores and the churchs themselves:

> *While the churches were gradually emptied by the wavering of faith, they were replaced in souls that were now empty by [Mouret's] emporium. Women came to him to spend their hours of idleness, the uneasy, trembling hours that they would once have spent in chapel: it was a necessary outlet for nervous passion, the revived struggle of a god against the husband, a constantly renewed cult of the body, with the divine afterlife of beauty. If he had closed his doors, there would have been a riot outside, the frantic cry of pious women denied the confessional and the altar.*

The rhetoric of chapels and cathedrals may seem excessive to the modern ear, but in a way those analogies make sense, given the historical context. We are accustomed now to grand buildings and other planned environments with lavish designs that are open to the general public: shopping malls, theme parks, office buildings, airport terminals. But to a member of the middle class in the nineteenth century, the experience of exploring a space as epic as Le Bon Marché would have been largely unprecedented; the closest equivalent would have been venturing into Notre Dame or St. Paul's. Grand palaces had existed for centuries, of course, but they were off-limits to ordinary people. Until Le Bon Marché, the only space that offered comparable grandeur to commoners was the church.

Zola's novel also introduced a narrative device still popular today in both Hollywood movies and the implicit narratives that we draw upon to make sense of social change. The main protagonist, Denise Baudu, is a young woman from the provinces who comes to the city to work at her uncle's small fabric store near Mouret's new emporium. She eventually leaves to work for Mouret, and much of the narrative involves her uncle's battle to stay in business, competing with Mouret's massive scale and ambition. This struggle would soon become an enduring refrain—the chain store overwhelming the indies and the mom-and-pops—and would form the narrative spine of movies like *The Shop Around the Corner* and *You've Got Mail.*

Le Bon Marché—and its contemporaries in London and New York—helped inaugurate many practices that became commonplace in the next century. The department stores largely killed off the practice of haggling over prices with vendors, replacing that model with high-volume, low-markup sales of goods with fixed prices. Department stores were among the very first institutions to experiment with charge cards that extended their customers a line of credit. (The modern credit card, honored by a wide range of establishments instead of a single store, wouldn't be invented until the 1950s.) The grand department stores were also the first large organizations focused on the services industry, managing thousands of employees and complex supply chains. Organizations of comparable complexity

Interior of Le Bon Marché, Paris

had emerged in the previous century, but they were inevitably focused on the problems of manufacturing or engineering: building railroads, for instance, or converting raw cotton into textiles. Until Le Bon Marché, the service industry could hardly be considered an industry at all, just a dispersed collection of small shopkeepers, restaurateurs, lawyers, and other miscellaneous trades. Today, three out of four American jobs are in the services sector; by some accounts, the service industry accounts for 80 percent of U.S. GDP. Up until the late 1800s, services seemed like an amusing ornament next to the big business of trade and manufacture, a folly of interest only to a small elite, like the cupola hovering above the rue de Sèvres. Le Bon Marché and its peers hinted, for the first time, that the service industry would soon become a cornerstone.

Women, of course, were at the epicenter of this new industry, as they had been during the dawn of the cotton revolution two centuries before. Increasingly, women were working in these new department stores, not just enjoying them as consumers. And just as cotton had unleashed a moral panic over the traitorous desires of the Calico Madam, Le Bon Marché triggered a similar crisis in Parisian society. In this case, it was not women *buying* luxury items that caused the outrage; it was the even more startling fact that women were *stealing* them.

Shortly after the arrival of *grands magasins* like Le Bon Marché, the Parisian authorities and other interested parties began noticing a marked uptick in the number of women caught stealing items from the stores, apparently motivated by some kind of deranged pleasure in the act. "I felt myself overcome little by little by a disorder that can only be compared to that of drunkenness, with the dizziness and excitation that are peculiar to it," one shoplifter testified. "I saw things as if through a cloud, everything stimulated my desire and assumed, for me, an extraordinary attraction. I felt myself swept along towards them and I grabbed hold of things without any outside and superior consideration intervening to hold me back. Moreover I took things at random, useless and worthless articles as well as useful and expensive articles. It was like a monomania of possession."

The term *kleptomania* had been around for several decades as an obscure medical diagnosis, after a wave of predominantly female thieves began pocketing goods from smaller stores in the opening decades of the century. Reading the early essays on the topic is like playing a greatest-hits collection of Victorian clichés: the shoplifter's malady is attributed to a "lesion of the will," to hysteria, menstruation, masturbation, and epilepsy. But the 1880s wave of shoplifting at Le Bon Marché and its competitors had elevated the crime to a central topic of conversation in the newspapers and drawing rooms of Paris, and in that elevation a new model for thinking about the human mind and its discontents began to take shape, however tentatively. The shoplifters were extensively studied by the scientific community. The psychiatrist Legrand du Saulle argued that the thefts "constitute a Parisian happening truly and completely contemporary, since they only date from the recent foundation and opening of the *grands magasins* themselves." He called the disorder, unforgettably, the department-store disease.

What made the wave of shoplifting particularly alarming—and also somewhat baffling—was the fact that so many of the culprits came from well-to-do households. The thefts were seemingly unmotivated by economic need. Another analysis by the criminologist Alexandre Lacassagne argued that "department store thefts have assumed in our day a real importance because of their growing number, the value and variety of goods stolen, [and] the quality of the persons committing these thefts." Later he remarked that "most of these kleptomaniacs are only arrested in the department stores. They steal there and nowhere else." Lacassagne used the department-store disease as a means of challenging the reigning orthodoxy of criminology, founded by the Italian Cesare Lombroso, which contended that criminal pathologies were due to inherited physiological defects. Because the syndrome only emerged with the new environment of the department store, and because the culprits themselves were so obviously of "good breeding," Lacassagne argued, the roots of the disease must be environmental.

Visible in these perplexed diagnoses is a new way of thinking about the psyche. Diseases of the mind did not have to be rooted in some biological deformity, as the phrenologists had contended; nor was it attributable to some abstract "lesion of the will"; nor was it tied to basic biological realities like menstruation or masturbation. Instead, the root cause of the disorder was to be found somewhere else: in the lived history of social and economic change. Modern life itself could make you sick.

Within a decade, Freud would reroute psychology back toward the eternal truths of Eros and Thanatos, the family romance, and the pleasure principle. But by the second half of the twentieth century, the kind of diagnosis that emerged with the department-store disease would become increasingly familiar. We now assume, correctly or not, that every new media experience is rewiring our brains in some fundamental way; today's disorders—attention deficit disorder, autism, teen violence—are regularly chalked up to the sensory overload of television, or video games, or social media. We take it for granted that the brain is shaped by the built environment that surrounds it, for better and for worse. That way of seeing the mind—and understanding its occasional defects—first came into view with the unlikely criminals of the department-store disease.

Architects had long constructed environments designed to trigger certain emotional responses in their visitors: think of the soaring interiors of the medieval cathedral, meant to inspire awe and wonder. Boucicaut's genius—some might call it an evil genius—lay in his realization that built environments could be created to *program* behavior, to make people want things they didn't know they needed, to feel desires they had not felt before they entered the wonderland of the *grand magasin*. But as the kleptomaniacs of the French upper classes made clear, even Boucicaut had no idea just how powerful that program would turn out to be.

Follow Route 35 southwest of Minneapolis to the suburban town of Edina, and take the exit onto West Sixty-Sixth Street, and you will eventually find a building complex floating like an island in a gray sea of parking, its exterior a jumbled mix of branded facades: GameStop, P. F. Chang's, AMC Cinemas. Sixty years ago, when it was first built, the exterior of the building was almost entirely deprived of ornament, the exact opposite of Le Bon Marché's caryatids and Roman gods. But the design of this structure would define its era every bit as much as Boucicaut's temple did, for this unremarkable building—practically indistinguishable from the shopping complexes and office parks that surround it on all sides—is Southdale Center, America's first mall.

Today malls have a mostly well-deserved reputation for being the ugly stepchild of consumer capitalism, but their intellectual lineage is more complex than most people realize. While it would come to epitomize the cultural wasteland of postwar suburbia, the shopping mall turns out to have been the brainchild of an avant-garde European socialist named Victor Gruen. Born in Vienna around the turn of the century, Gruen grew up, as his biographer M. Jeffrey Hardwick put it, "in the dying embers of [Vienna's] vibrant, aesthetic life." He studied architecture at the Vienna Academy of Fine Arts, working under the socialist urban planners then in vogue, and performing in cabarets at night. He built up a fledgling practice designing storefronts on the fashionable streets of Vienna, not unlike the original merchants in Ludgate Hill so many years before, and he designed—but never built—one large-scale project for public housing, which he dubbed "The People's Palace." Like many left-wing Jewish intellectuals, Gruen fled to the United States as the

Victor Gruen

Nazis began marching across Europe. (He left Vienna the same week Sigmund Freud did.) He arrived in the United States not speaking a word of English, but by 1939 he was performing with a theater troupe in Manhattan and designing boutiques on Fifth Avenue. He developed a signature style in the shop designs, with open-air arcade entrances flanked by giant plate-glass displays arrayed with goods. The new storefronts delighted consumers and merchants alike, though critics like Lewis Mumford grumbled that the facades captured their customers the way "a pitcher plant captures flies or an old-style mousetrap catches mice." During the 1940s, Gruen's design practice boomed; he built dozens of department stores across the country. Echoing Le Corbusier's famous line about a house being a "machine for living," Gruen began calling his store environments "machines for selling."

Yet Gruen never fully left his Viennese radical upbringing and its faith in the potential of large-scale planned communities. He hated the noisy, crass commercialism of unregulated spaces. He had an urbane European's disdain for American suburbia. In the late 1950s, Gruen gave a speech in which he denounced the banal landscapes of the postwar suburbs, calling them "avenues of horror . . . flanked by the greatest collection of vulgarity—billboards, motels, gas stations, shanties, car lots, miscellaneous industrial equipment, hot dog stands, wayside stores—ever collected by mankind." Gruen was a complicated mix: a socialist who despised the aesthetics of unchecked capitalism, who nevertheless designed department stores for a living. His job description was almost as hard to classify as his values. A mix of architect, urban planner, and interior decorator, Gruen eventually began calling himself an "environmental designer," a phrase that Boucicaut would have understood in an instant.

In the late 1940s and early 1950s, Gruen began exploring more ambitious designs that would incorporate multiple stores and other public spaces. He designed a hugely successful open-air shopping plaza in the suburbs of Detroit called Northland Center. But it was in 1956 that Gruen completed work on Southdale Center, which would become his most famous—and, to some, notorious—project. Gruen designed Southdale as a two-level structure linked by opposing escalators, featuring a few dozen stores arrayed around a shared courtyard, protected from the harsh Minneapolis weather by a roof. Gruen modeled the building after the European arcades that had flourished in Vienna and other cities in the early nineteenth century. But to modern eyes, the reference to European urbanity is lost: Southdale Center is, inescapably, a shopping mall, the first of its kind to be built.

Southdale was an immediate hit, attracting almost as much cov-

Southdale Center, America's first mall

erage and hyperbolic praise as Walt Disney's reinvention of the amusement park. "The strikingly handsome and colorful center is constantly crowded," *Fortune* announced. "The sparkling lights and bright colors provide a continuous invitation to look up ahead, to stroll on to the next store, and to buy." Ecstatic photo essays appeared everywhere from *Life* to *Time* to *Architectural Forum*, the latter of which described Southdale as "more like downtown than downtown itself." Most commentators focused on the expansive courtyard space, which Gruen had dubbed the "Garden Court of Perpetual Spring," where shoppers could enjoy sculptures, children's carnivals, cafés, eucalyptus and magnolia trees, birdcages, and dozens of other diversions. Interestingly, one of the few dissenting voices came from Frank Lloyd Wright, then nearing ninety years of age, who complained that the garden court "has all the evils of a village street and none of its charms."

Gruen's design for Southdale would become the single most influential new building archetype of the postwar era. Just as Louis Sullivan's original skyscrapers had defined the urban skylines of the first half of the twentieth century, Gruen's shopping mall proliferated around the globe, first in suburban American towns newly populated by white flight émigrés from metropolitan centers. Shopping meccas like L.A.'s Beverly Center became cultural landmarks, and the default leisure activity of hanging at the mall would define an entire generation of "Valley girls." But as mall culture went global, Gruen's design became increasingly prominent in the downtown centers of new megacities. Originally conceived as a way to escape the harsh winters of Minnesota, Gruen's enclosed public space accelerated the mass migration to desert or tropical climates made possible by the invention of air-conditioning. Today, the ten largest shopping malls in the world are all located in non-U.S. or European countries with tropical or desert climates, such as China, the Philippines, Iran, and Thailand. And while the mall itself would expand in scale prodigiously—a mall in Dubai has more than one thousand stores spread out over more than five million square feet of real estate—the basic template of Gruen's design would remain constant: two to three floors of shops surrounding an enclosed courtyard, connected by escalators.

But there is a tragic irony behind Gruen's seemingly massive success. The mall itself was only a small part of Gruen's design for Southdale and its descendants. Gruen's real vision was a for a dense, mixed-use, pedestrian-based urban center, with residential apartments, schools, medical centers, outdoor parks, and office buildings. He later expanded his vision of the new city in an eclectic series of planning briefs, speeches, and essays, culminating in a book called *The Heart of Our Cities.* The spectacle of the mall courtyard, and its pedestrian convenience, was for Gruen a way to smuggle European metropolitan values into a barbaric American suburban wasteland. According to Gruen's original design, as Malcolm Gladwell writes, "Southdale was not a suburban alternative to downtown Minneapolis. It was the Minneapolis downtown you would get if you started over and corrected all the mistakes that were made the first time around." Even the ultimate defender of traditional downtown sidewalks, Jane Jacobs, was smitten by Gruen's designs. Describing an ambitious plan for a new Fort Worth that Gruen had developed but never built, Jacobs wrote, "The service done by the Fort Worth plan is of incalculable value, [and will] set in motion new ideas about the function of the city and the way people use the city."

Yet developers never took to Gruen's larger vision: instead of surrounding the shopping center with high-density, mixed-use development, they surrounded it with parking lots. They replaced Gruen's courtyard carnivalesque with food courts. Communities did blossom around the new malls, but they were largely uncoordinated developments of low-density single-family homes. The suburbs had always been safer and more family-friendly than city centers, but Gruen's shopping mall held out the promise that they might also be *exciting* in the way that Fifth Avenue or the Miracle Mile was exciting. Of course, suburbanization had many winds in its sails, but Gruen's shopping mall was undeniably one of the strongest. Southdale was going to be the antidote to suburban sprawl. Instead it became an amplifier.

And here is where the story turns dark. The mall didn't just help create the modern, postwar suburb, it also helped undermine the prewar city. The mass exodus from urban centers in Detroit and Minneapolis and Brooklyn and dozens of other U.S. cities precipitated the urban crises of the 1960s. No affliction is more devastating to the life cycles of a big city than sudden population loss. Race riots, exploding crime rates, abandoned neighborhoods, and budget crises convinced many reasonable Americans that big cities were either confronting total collapse, or were doomed to live in a permanent state of anarchy. Those eulogies for the city turned out to be premature, and today it is the old downtown that is drawing shoppers and strollers and flaneurs away from the mall. But for a few decades there, it was touch and go. It is no accident that cities like Detroit that are still struggling to climb back from the urban collapse of the 1960s were also the ones where Gruen's designs first took root. Even his cherished Vienna, he discovered on returning to the city in the early 1970s, had been threatened by what a he called a "giant shopping machine" built on the outskirts of the city, imperiling the small, independent stores within the city center. Gruen would eventually renounce his creation, or at least the distorted version of it that the mall developers had implemented: "I refuse to pay alimony," he proclaimed, "for these bastard developments."

As the shopping-center developers were happily ignoring Gruen's plans for reinventing the city, his ideas nonetheless managed to attract one devoted fan who had the financial resources to put them into action: Walt Disney. The 1955 launch of Disneyland had been a staggering success for Disney. No one had ever built a theme park with such attention to detail, a fantasy space that enveloped the visitor, a perfect cocoon of wonder and amusement. But the triumph of the planned environment inside the park itself created a kind of opposing reaction in the acres outside, which were swiftly converted from orange groves into cheap motels, gas stations, and billboards. Disney grew increasingly repulsed by the blight of highway sprawl that surrounded his crown jewel. And so he began plotting to construct a second-generation project in which he could control the whole environment—not just the theme park but the entire community around it.

In what was going to be the ultimate act of imagineering, Disney planned to design an entire functioning city from scratch, one that would reinvent almost every single element of the modern urban experience. He had dubbed it EPCOT, short for Experimental Prototype Community of Tomorrow. While the Disney Corporation would eventually build a future-themed amusement park called EPCOT in Orlando, it had virtually nothing to do with Disney's vision for EPCOT, which would have been a true community with full-time residents, not another tourist attraction. In his career, Disney had radically reinvented multiple forms of entertainment, from animated features to amusement parks. But his final act was going to be even more ambitious: reinventing urban life itself.

The legend that has accumulated around Disney's epic goals for EPCOT largely fits with the historical facts. In the 1966 film introducing Disney World, Disney spends almost no time discussing the amusement-park component of the project (what would eventually become the Magic Kingdom). Instead, he focuses extensively on his "city of tomorrow," showing prototypes and sketches that look strikingly like the futurist cityscapes imagined by Le Corbusier almost fifty years before. But the legend of Walt's avant-garde urban planning has a strange twist to it, one that has received less coverage over the years. In preparation for the EPCOT project, Disney went on a national tour of visionary, cutting-edge experiments in planning and community design, seeking inspiration for his own radical new city. What were the primary sacred sites for such a pilgrimage? Two shopping malls on the East Coast, as well as a new Neiman Marcus department store in Texas.

The gap between Disney's ambition and his prior art seems a little baffling to us now, like someone setting out to reinvent the literary novel by studying the Great Works of Nicholas Sparks. But the fact is, Disney's expedition made a certain kind of sense: the shopping mall was, in the early 1960s, a new kind of planned environment, one that would have a tremendous impact on social organization in the decades to come. Like it or not, it was a crystal ball. If you wanted to understand something important about the way human beings would live in the year 2000, studying shopping malls was probably about as good a clue as anything going in the early 1960s.

During his exploratory research, Disney fell under the spell of Gruen. Gruen had included some kind words about the planned environment of Disneyland in *The Heart of Our Cities*, and predictably shared Disney's contempt for the sprawling "avenues of horror" that had proliferated around the theme park in Anaheim. And so when Disney decided to buy a vast swath of swampland in central Florida and build from scratch a "Progress City," as he originally called it, Gruen was the perfect patron saint for the project.

In 1966, Disney set up a Skunk Works operation on their Burbank lot, in a lofty space that was quickly dubbed "The Florida Room," where a team of imagineering urban planners labored over mock-ups of their new city. Disney apparently kept a copy of *The Heart of Our Cities* on his desk. For months, the existence of the room was top secret, even to Disney insiders. (The secrecy derived in part from the fact that Disney had to purchase most of the Florida land parcels anonymously to keep the price low.) But once word leaked of the Orlando site, Disney decided to open the doors to the Florida Room and to the project it was incubating. In the late summer of 1966, he made a thirty-minute film introducing Disney World that featured dazzling footage of the twenty-foot maps covering the Florida Room's oversized walls. (Draftsmen are seen climbing ladders, presumably adding some intoxicating new detail to one of the maps.) A kind of aura has developed around the film among Disney cognoscenti, because it turned out to be the last film that Walt Disney ever made. (He was already terminally ill with cancer during the filming.) But it is more interesting to us today as a glimpse of what Disney might have built had he not died.

The film makes it abundantly clear how central the EPCOT project was to Disney. Standing in front of an oversized map of the entire project, Disney spoke of his ambitions in his usual avuncular manner:

> The most exciting, by far the most important part of our Florida project—in fact, the heart of everything we'll be doing in Disney World—will be our experimental prototype city of tomorrow . . . EPCOT will take its cue from the new ideas and new technologies that are now emerging from the creative centers of American industry. It will be a community of tomorrow that will never be completed, but will always be introducing, and testing, and demonstrating new materials and new systems. And EPCOT will always be a showcase to the world of the ingenuity and imagination of American free enterprise. I don't believe there is a challenge anywhere in the world that's more important to people everywhere than finding solutions to the problems of our cities.

The first thing that should be said about EPCOT is that, like Gruen's original plan for Northdale, it was going to be an entire community oriented around a mall. "Most important, this entire fifty acres of city streets and buildings will be completely enclosed," a narrator explains in the 1966 film. "In this climate-controlled environment, shoppers, theatergoers, and people just out for a stroll will enjoy ideal weather conditions, protected day and night from rain, heat and cold, and humidity."

The vision of a radical new model of urbanism with a shopping mall at its core is enough to produce snickers from the modern urbanist, for good reason. Yet the EPCOT plan had a complexity to it that we should not forget, a complexity that no doubt derived from the contradictions in Gruen's philosophy. It was, for starters, pro-

Early models for EPCOT's World Showcase. The designs reflect traces of Disney's original plan to build a centrally organized pedestrian city.

foundly anti-automobile. At the center of the city was a zone that Gruen had come to call the Pedshed, defined by the "desirable walking distance" of an average citizen. Cars would be banned from the entire Pedshed area. As one moved away from the central core, new modes of transportation would appear, each appropriate to the distance required to get EPCOT residents downtown: the electric "people movers" that now take tourists around Disney World's Tomorrowland would shuttle residents from the high-density apartments to the commercial core; longer trips to and from the lower-density residential developments and industrial parks at the edge of the city would be conducted via monorail. Just as in Disney's theme parks, all supply and service vehicles would be routed below the city through a network of underground tunnels. In the film, the narrator happily explains that EPCOT residents would use their cars only on "weekend pleasure trips."

The tragic contradictions of Gruen's life run through the plan for EPCOT as well: watching Disney's film, you catch a fleeting glimpse of an alternate version of the recent past, where the pedestrian mall—"more like downtown than downtown itself"—inspires a new vision of urban life that rejects the tyranny of the automobile and ushers in a new era of mass-transit innovation. (Just imagine the impact on climate change if we'd had thirty years of using our automobiles only for weekend pleasure trips.) But, of course, that alternate past didn't happen. Instead, the mall triggered decades of suburban ascendancy, and the Walt Disney Corporation turned EPCOT into yet another theme park, with its bizarre and sad hybrid of Buckminster Fuller futurism and It's-a-Small-World globalism.

Why weren't progress cities built? The easiest way to dismiss the Gruen/EPCOT vision is to focus on the centrality of the mall itself. Now that mall culture is in decline—in the United States and Europe at least—we understand that the overly programmed nature of the mall environment ended up being its fatal flaw. As always, play is driven by surprise and novelty, just as it was when those London ladies first encountered the lavish shopfronts of Ludgate Hill, or when the Parisian kleptomaniacs first wandered into the wonderland of Le Bon Marché. One of the reasons the critics raved over Gruen's original Southdale Center was the simple fact that no one had seen a space like that before, particularly in suburban Minnesota. But as the developers standardized Gruen's original plan, and as the big chain stores grew more powerful, malls became interchangeable: a characterless cocoon of J.Crew and the Body Shop and Bloomingdale's. They were not quite "avenues of horror" but something equally soulless: avenues of sameness. Eventually, our appetite for novelty and surprise overcame the convenience and ubiquity of mall culture, and people began turning back to the old downtowns. Those downtowns were dirty and crowded and open to the elements, but they were also unpredictable and unique and fun in a way that the mall could never be.

Disney and Gruen wanted the energy and vitality and surprise of the big city, without all of the hassle. It turns out that a little bit of hassle is the price you pay for energy and vitality. But I suspect the mall at the epicenter of Southdale and EPCOT is too distracting a scapegoat: dismissing EPCOT as a crowning moment in the history of suburbanization—the city of the future is built around a mall!—diverts the eye from the other elements of the plan that actually have value. The fact that Jane Jacobs, who had an intense antipathy to top-down planners, saw merit in the Gruen model should tell us something. It would be fitting, in a way, if some new model of urban organization emerged out of a shop-window designer's original vision, given the roots of the industrial city in the lavish displays of the London shops. Routing services belowground; clearing out automobiles from entire downtowns; building mixed-use dense housing in suburban regions; creating distinct mass-transit options to fit the scale of the average trip—these are all provocative ideas that have been explored separately in many communities around the world. But, to this day, no one has built a true Progress City, which means we have no real idea how transformative it might be to see all these ideas deployed simultaneously. Mall or no mall, perhaps it's time we tried.

2

Music

The Machine That Plays by Itself

Roughly forty-three thousand years ago, a young cave bear died in the rolling hills on the northwest border of modern-day Slovenia. A thousand miles away, and a thousand years later, a mammoth died in the forests above the river Blau, near the southern edge of modern-day Germany. Within a few years of the mammoth's demise, a griffon vulture also perished in the same vicinity. Five thousand years after that, a swan and another mammoth died nearby.

We know almost nothing about how these different animals met their deaths. They may have been hunted by Neanderthals or modern humans; they may have died of natural causes; they may have been killed by other predators. Like almost every creature from the Paleolithic era, the stories behind their lives (and deaths) are a mystery to us, lost to the unreconstructible past. But these different creatures—dispersed across both time and space—did share one remarkable posthumous fate. After their flesh had been consumed by carnivores or bacteria, a bone from each of their skeletons was meticulously crafted by human hands into a flute.

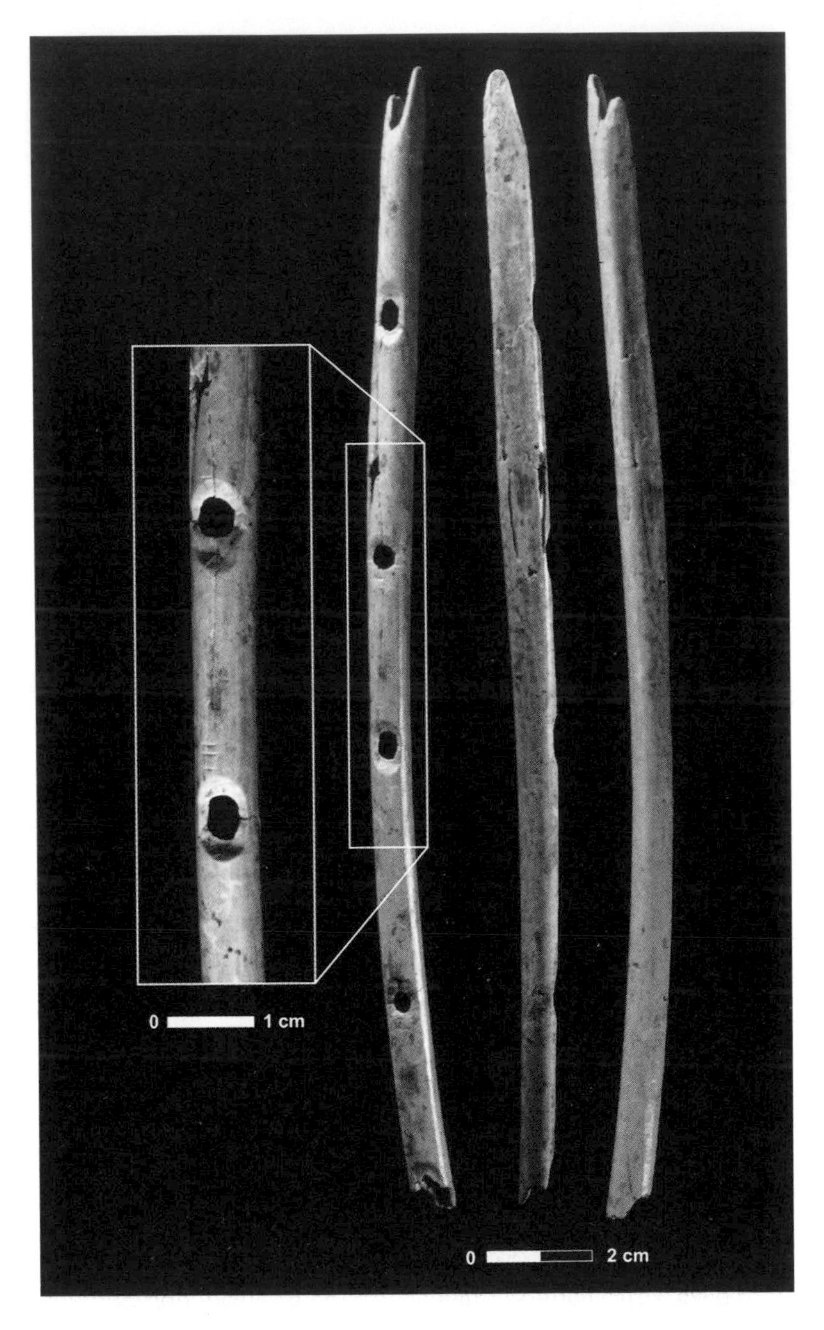

A bone flute from around 33,000 BCE

Bone flutes are among the oldest known artifacts of human technological ingenuity. The Slovenian and German flutes date back to the very origins of art; the caves where some of them were found also featured drawings of animals and human forms on their walls, suggesting the tantalizing possibility that our ancestors gathered in the firelit caverns to watch images flicker on the stone walls, accompanied by music. But musical technology is likely far older than the Paleolithic. The Slovenian and German flutes survived because they were made of bone, but many indigenous tribes in modern times construct flutes or drums out of reeds and animal skin, materials unlikely to survive tens of thousands of years. Many archeologists believe that our ancestors have been building drums for at least a hundred thousand years, making musical technology almost as old as technology designed for hunting or temperature regulation.

This chronology is one of the great puzzles of early human history. It seems to be jumping more than a few levels in the hierarchy of needs to go directly from spearheads and clothing to the invention of wind instruments. Aeons before early humans started to imagine writing or agriculture, they were crafting tools for making music. This seems particularly puzzling because music is the most abstract of the arts. Paintings represent the inhabitants of the world that our eyes naturally perceive: animals, plants, landscapes, other people. Architecture gives us shelter. Stories follow the arc of events that make up a human life. But music has no obvious referent beyond a vague association with the chirps and trills of birdsong. No one likes a hit record because it sounds like the natural world. We like music because it sounds like music—because it sounds *different* from the unstructured noise of the natural world. And what sounds like music is much closer to the abstracted symmetries of math than any experience a hunter-gatherer would have had a hundred thousand years ago.

A brief lesson in the physics of sound should help underscore the strangeness of the archeological record here. Some of the bone flutes recovered from Paleolithic cave sites are intact enough that they can be played, and in many cases, researchers have found that the finger holes carved into the bones are spaced in such a way that they can produce musical intervals that we now call perfect fourths and fifths. (In the terms of Western music, these would be F and G in the key of C.) Fourths and fifths not only make up the harmonic backbone of almost every popular song in the modern canon, they are also some of the most ubiquitous intervals in the world's many musical systems. Though some ancient tonal systems, like Balinese gamelan music, evolved without fourths and fifths, only the octave is more common. Musicologists now understand the physics behind these intervals, why they seem to trigger such an interesting response in the human ear. An octave—two notes exactly twelve steps apart from each other on a piano keyboard—exhibits a precise 2:1 ratio in the wave forms it produces. If you play a high C on a guitar, the string will vibrate exactly two times for every single vibration the low C string generates. That synchronization—which also occurs with the harmonics or overtones that give an instrument its particular timbre—creates a vivid impression of consonance in the ear, the sound of those two wave forms snapping into alignment every other cycle. The perfect fourth and fifth have comparably even ratios: a fourth is 4:3, while a fifth is 3:2. If you play a C and a G note together, the higher G strings will vibrate three times for every two vibrations of the C. By contrast, a C and F# played together create the most dissonant interval in the Western scale—the notorious tritone, once called the "devil's interval"—with a ratio of 43:32.

The existence of these ratios has been known since the days of ancient Greece; the tuning system that features them is often called Pythagorean tuning, after the Greek mathematician who, legend has it, first identified them. (Today, the average seventh grader knows Pythagoras for his triangles, but his ratios are the cornerstone of every pop song on Spotify.) The study of musical ratios marked one of the very first moments in the history of knowledge where mathematical descriptions productively explained natural phenomena. In fact, the success of these mathematical explanations of music triggered a two-thousand-year pursuit of similar cosmological ratios in the movements of the sun and planets in the sky—the famous "music of the spheres" that would inspire Kepler and so many others.

Wave forms, integer ratios, overtones—none of these concepts were available to our ancestors in the Upper Paleolithic. And yet, for some bizarre reason, they went to great lengths to build tools that could conjure these mathematical patterns out of the simple act of exhaling. Put yourself in that Slovenian cave forty thousand years ago: you've mastered fire, built simple tools for hunting, learned how to craft garments from animal skins to keep yourself warm in the winter. An entire universe of further innovation lies in front of you. What would you choose to invent next? It seems preposterous that you would turn to crafting a tool that created vibrations in air molecules that synchronized at a perfect 3:2 ratio when played together. Yet that is exactly what our ancestors did.

Why were early humans so intent on producing acoustic waves that traveled in such clean integer ratios? The fact that they made musical instruments without understanding the fundamentals of acoustic theory is not so puzzling, of course. There were obviously many cases where evolution had steered them toward objectives whose underlying nature was not at all understood; the Paleolithic hunter-gatherers sought out—and built tools to acquire—the tastes of sugars and fats without knowing anything about the molecular chemistry of carbohydrates and lipids. But sugars and fats are essential components of our diets; we die if we don't eat enough of them. Listening to a 4:3 ratio doesn't bestow the same obvious benefits. Our hominid relatives had survived for millions of years without hearing them at all. "Without music, life would be a mistake," Nietzsche famously proclaimed. But there would still be *life*. Without sex, or water, or proteins, the human race would cease to exist. Without perfect fourths, we'd be robbed of the bass hook from "Under Pressure"—and just about every piece of music written in the past few centuries—but we'd survive as a species.

I suspect music first emerged not with a need but with a difference: an unusually resonant sound happened to emerge out of the structure of some hollow object—a reed or a bone—creating a tone just different enough from the ordinary cacophony of the world that the ear took note. The sound wasn't meaningful yet, or laden with the kind of emotional overtones that humans now associate with music. It was just new. And like the unusual shade of Tyrian purple, because the sound was new, it was interesting, worth repeating, worth tinkering with. As these early instruments began to be capable of triggering octaves when played as an ensemble, it may be that our distant ancestors found the sound particularly evocative because male and female voices are, on average, roughly an octave apart, and so the strange consonance of the proto-flute seemed to echo the sound of a conversation between a man and a woman. Perhaps chanting in harmony predated the first instruments, and so when the random variation of evolution happened to concoct a bone or a stalk that by sheer coincidence generated an octave, and we in turn stumbled across that bone or stalk, we found the echoes so alluring that we set about engineering the effect ourselves.

Because music has such a long history in human society, some scientists believe that an appetite for song is part of the genetic heritage of *Homo sapiens*, that our brains evolved an interest in musical sounds the way it evolved color perception or the ability to recognize faces. The question of whether music is a cultural invention or an evolutionary adaptation has been a contentious one in the last decade or so, a debate initially triggered by Steven Pinker's best-selling manifesto of evolutionary psychology, *How the Mind Works*. Pinker is famous for seeing the mind as a kind of toolbox with a set of specific attributes shaped by the evolutionary pressures of our ancestral environments. But music he considers to be a cultural hack, designed to trigger circuits in the brain that evolved for more pressing tasks. In one of the book's most controversial passages, he compared music to strawberry cheesecake:

> *We enjoy strawberry cheesecake, but not because we evolved a taste for it. We evolved circuits that gave us trickles of enjoyment from the sweet taste of ripe fruit, the creamy mouth feel of fats and oils from nuts and meat, and the coolness of fresh water. Cheesecake packs a sensual wallop unlike anything in the natural world because it is a brew of megadoses of agreeable stimuli which we concocted for the express purpose of pressing our pleasure buttons . . . music is auditory cheesecake, an exquisite confection crafted to tickle the sensitive spots of at least six of our mental faculties.*

Something about that cheesecake metaphor did not sit well with other scientists, and in the years that followed, many argued that the taste for music must have had some direct adaptive value, given the prominence of musical instruments in the early archeological record—and the ubiquity of music across all human societies. Some believe that musical chanting may have predated language itself, that words and sentences evolved out of the prelinguistic communication of harmony and rhythm. The researchers who discovered one set of bone flutes in Germany argue that music may have played a key role in cementing social bonds among early humans: "The presence of music in the lives of early Upper Paleolithic peoples did not directly produce a more effective subsistence economy and greater reproductive fitness," they write. "Viewed, however, in a broader behavioral context, early Upper Palaeolithic music could have contributed to the maintenance of larger social networks, and thereby perhaps have helped facilitate the demographic and territorial expansion of modern humans relative to culturally more conservative and demographically more isolated Neanderthal populations." Others take the sexual conquests of modern musicians as a sign that musical talent may be a trait encouraged by sexual selection: having a gift for song didn't make you more likely to survive the challenges of life in the Upper Paleolithic, but it did make you more likely to reproduce your genes.

One premise unites both sides in this debate: that music "presses our pleasure buttons," as Pinker describes it. Yet there is something too simple in describing our appetite for music in this way. Sugar and opiates, to give just two examples, press pleasure buttons in the brain in a relatively straightforward fashion. Given a taste of one, we instinctively return for more of the same, like those legendary lab rats endlessly pressing the lever for more stimulants. And we put our ingenuity to work concocting ever-more-efficient delivery mechanisms for these forms of pleasure: we refine opium into heroin; we start selling soda in Big Gulp containers. But music—like the patterns and colors unleashed by the fashion revolution—appears to resonate with our pleasure centers at more of an oblique angle. The pleasure in hearing those captivating sounds doesn't just establish a demand for more of the same. Instead, music seems to send us out on a quest for new experiences: more of the same, but different.

Wherever you fall on the evolutionary question, music confronts us with one undeniable paradox: this most abstract and ethereal of entertainments—conjured up out of invisible symmetries of air molecules vibrating—has a longer history of technological innovation than any other form of art. Since tones generated by that first bone flute resonated in our ears, we've been chasing new sounds, new timbres, new harmonies. And that pursuit led to countless technological breakthroughs that shaped modern life in entirely nonmusical ways.

The pursuit of novelty recurs again and again in the history of play. One way to imagine it is that evolution has given us two kinds of pleasure buttons. The first is an all-hands-on-deck kind of button: we need food; we need warmth; we need offspring. If we don't have those things, we will die, or our genes will not be passed on to the next generation. So there are "pleasure buttons" associated with satisfying those needs: the pleasures of sex, or eating proteins and carbohydrates. But there are other, less urgent pleasures, like the sound of air being blown through a vulture bone. We don't *need* to hear that sound in any existential sense, but nonetheless something about it captures our attention, prods us to seek the experience out in future environments. But at the same time, something about it compels us to vary the experience. The pure pleasure buttons in the brain, like the endorphin system, don't compel you to seek out anything other than increasing amounts of endorphins. In fact, the pleasure associated with them is so powerful that most people who get addicted to artificial versions, like heroin or OxyContin, lose interest in other experiences altogether. The pull of opiates is centripetal; most heroin victims die alone for a reason. But music, like other similar forms of play, is a push: it propels you to seek out new twists.

That exploratory, expansive drive is what separates delight from demand: when we are in play mode, we are open to new surprises, while our base appetites focus the mind on the urgent needs of staying alive. Understanding that distinction is critical to understanding why play—despite its seemingly frivolous veneer—has led to so many important discoveries and innovations. The question of why the *Homo sapiens* brain possesses this strange hankering for play and surprise is a fascinating one, and I will return to it in the final pages of this book. But for now, we need to establish just how far those playful explorations took us.

The bone flutes must have sounded enchanting to the early humans of the Upper Paleolithic. But they were just the beginning.

The Banu Musa—those brilliant toy designers from the Islamic golden age—earned a permanent place in the pantheon of engineering and robotics with their *Book of Ingenious Devices*. And yet the brothers omitted from that collection what may have been their most ingenious device of all, a machine that would introduce one of the most important concepts of the digital age more than a thousand years before the first computers were built. Evidence of their design lies in a separate treatise, transcribed by a scholar in the twelfth century and discovered a hundred years ago in a library at Three Moons College in Syria. Like most of the Banu Musa's work, the document is an intensely technical how-to guide for building a machine with hundreds of distinct components. But the title has a captivating clarity to it: "The Instrument Which Plays by Itself."

"We wish to explain," the brothers announce in the opening lines, "how an instrument is made which plays by itself continuously in whatever melody we wish, sometimes in a slow rhythm and sometimes in a quick rhythm, and also that we may change from melody to melody when we so desire." The instrument in question was a hydraulic, or water-powered, organ, similar in design to organs built by the Greeks and Romans centuries before. Yet the Banu Masu device had one crucial feature that no instrument designer had ever implemented. The notes played by the organ were not triggered by human fingers on a keyboard. Instead, they were triggered by what came to be known as a pinned cylinder—a barrel with small "teeth," as the brothers called them, irregularly distributed across its surface. As the barrel rotated, those teeth activated a series of levers that opened and closed the pipes of the organ. Different patterns of "teeth" produced different melodies as the air was allowed to flow through the pipes in unique sequences. The brothers explained how a melody could be encoded onto these cylinders by capturing the notes played by a live musician on a rotating drum covered by black wax, strongly reminiscent of the phonographic technology that wouldn't be invented for another thousand years. (The Banu Musa system recorded the specific notes played, not sound waves, however.) The marks etched in the wax could then be translated into the teeth of the cylinder. Anticipating the slang that eventually developed for the record industry, the Banu Musa described this process as "cutting" a melody. In a fitting echo of the musical innovations

Reconstruction of the Banu Musa's self-playing music automaton

that preceded them, the Banu Musa even included a description of how their instrument could be embedded inside an automaton, creating the illusion that the robot musician was playing the encoded melody on a flute.

The result was not just an instrument that played itself, as marvelous as that must have been. The Banu Musa were masters of automation, to be sure, but humans had been tinkering with the idea of making machines move in lifelike ways since the days of Plato. Animated peacocks, water clocks, robotic dancers—all these contraptions were engineering marvels, but they also shared a fundamental limitation. They were locked in a finite routine of movements. But the instrument that played itself was not restrained in the same way. You could dream up new melodies for it, instruct it to produce new patterns of sound. The water clocks the Banu Musa built were automated. But their "instrument" was endowed with a higher-level property. It was *programmable.*

Conceptually, this was a massive leap forward: machines designed specifically to be open-ended in their functionality, machines controlled by *code* and not just mechanics. A direct line of logic connects the "Instrument Which Plays by Itself" to the Turing machines that have so transformed life in the modern age. You can think of the instrument as the moment where the Manichean divide between hardware and software first opened up. An invention that itself makes invention easier, faster, more receptive to trial and error. A virtual machine.

Yet it must have been hard to see (or hear) its significance at the time. To the untutored spectator, it might have paled beside the animated peacocks. So the organ sometimes played one sequence of notes, and sometimes played other tunes—how big a deal could that be? The textual evidence suggests that the Banu Musa were onto the significance of the advance; they devoted many pages in their assembly manual to discussing the technique for "cutting" new cylinders and for modifying the playback of existing melodies. But they couldn't have grasped where that step change would eventually lead.

Something about the concept of programmable music continued to attract the attention of engineers and instrument designers during the centuries that followed. Behind timepieces, music boxes were some of the most advanced works of mechanical engineering in the sixteenth and seventeenth centuries; almost without exception, these devices employed the pinned cylinder devised by the Banu Musa to program the melody and chords. Like the automata that succeeded them, these contraptions were meant to delight and amuse the elite, not compete with serious musical performances. They were playthings, novelties—but, through that play, a larger and more revolutionary idea was slowly taking shape.

The next milestone in the evolution of that idea would have been visible to any curious-minded Parisian in the late 1730s. Wandering into the reception room of the Hôtel de Longueville in Paris, on the site now occupied by part of the Tuileries, one would have seen, amid the ornate furnishings, a life-sized shepherd crouching on a pedestal, playing a flute. While the shepherd was in fact a machine, it played the flute the way any human would: by blowing air through the mouthpiece in varying rhythms, while covering and uncovering air holes on the body of the flute itself. Hidden inside the pedestal, air pumps and crankshafts controlled the pressure of the air released through the shepherd's mouth, along with the movement of the automaton's fingers. A pinned cylinder controlled the volume and sequence of the notes played; a rotating collection of cylinders allowed the shepherd to play twelve distinct songs.

The flute player was the creation of Jacques de Vaucanson, the French automaton designer now most famous for his "digesting duck." (The duck would appear at the Hôtel de Longueville on a pedestal next to the flute player the following year.) Vaucanson was the first of the automaton designers to focus on creating truly lifelike behavior in his machines, with movements that were predicated on careful anatomical study. After a decade of desultory work and travel displaying a handful of early designs, Vaucanson had attracted the patronage of a Parisian gentleman named Jean Marguin. With sufficient economic backing for the first time in his life, Vaucanson set out to design a machine that lived up to his ambitions: technology so advanced it would simulate the breath of life itself. Without realizing it, he was retracing the steps that the Paleolithic toolmakers had followed forty thousand years before. He took the most advanced engineering knowledge the world had ever seen, and used it to create 4:3 ratios in sound waves by blowing air through the holes carved in a hollow tube.

Jacques de Vaucanson

The bio-mimicry of the flute player's finger work and breathing would have impressed the Banu Musa, but the fundamental principles behind Vaucanson's design were essentially the same as the "Instrument Which Plays by Itself." The "programming" that controlled the machine's behavior came from the patterns of teeth on a rotating cylinder, and the power of that programmability was harnessed to play music. We should not overlook the strange nature of the technological history here: for eight hundred years, humans had possessed the protean resource of programmability, and over that time they had used that resource exclusively to generate pleasing patterns of sound waves in the air. (And, in the case of the flute player, to accompany those sound waves with physical movements that mimicked the behavior of human musicians.) Think of all the ways that the world has been transformed by software, by machines whose behavior can be sculpted and reimagined by new instruction sets. For almost a thousand years, we had that meta-tool in our collective toolbox, and we did nothing with it other than play music.

Vaucanson's flute player, however, would lead us out of that functional cul-de-sac. Designing the programmable cylinders that brought the musical shepherd to life suggested another application to Vaucanson, one that had far more commercial promise than showcasing androids in hotel lobbies. If you could use pinned cylinders to trigger complex patterns of sound waves, Vaucanson thought, why couldn't you use the same system to trigger complex patterns of *color*? If you could build a machine with enough mechanical dexterity to play a flute, why couldn't you build a similar machine to weave a pattern out of silk?

By 1741, Vaucanson's scandalous triumph with the digesting duck had made him a minor celebrity in France, acclaimed for his showmanship as much as for his engineering; he parlayed that success into a royal appointment advising Louis XV on plans to revitalize the French weaving industry, which was widely considered to have lagged behind its more technically ambitious competitors on the other side of the Channel. He toured the preindustrial looms of the country and began making plans for a machine that would do to fabric what the Banu Masu had done to melody. Instead of opening airholes in an organ or moving an android's fingers, Vaucanson's loom would control an array of hooks and needles that switched between different warp threads according to instructions encoded in a pinned cylinder. The same machine could be taught to weave a vast set of potential patterns out of silk. For the first time, programming would escape the boundaries of song.

That, at least, was the theory. In practice, the machine that Vaucanson ultimately built was hamstrung by the difficulty of translating the patterns onto the pinned cylinders, which were expensive to "cut." The patterns themselves had to adhere to repetitive designs, since the pattern looped with each rotation of the cylinder. Though several working prototypes were built, the machine never found a home in the French textile industry. But one of those prototypes survived long enough to find its way into the collection of the Conservatoire des Arts et Métiers, an institute formed in the early days of the French Revolution, more than a decade after Vaucanson's death. In 1803, an ambitious inventor from Lyon named Joseph-Marie Jacquard made a pilgrimage to the conservatoire to inspect Vaucanson's automated loom. Recognizing both the genius and the limitations of the pinned cylinder, Jacquard hit upon the idea of using a sequence of cards punched with holes to program the loom. In Jacquard's design, small rods, each attached to an individual thread, pressed against the punch cards; if they encountered the card's surface, the thread remained stationary. But if the rod pushed through a hole in the card, the corresponding warp thread would be woven into the fabric. It was, in its way, a kind of binary system, the holes in the cards reflecting on-off states for each of the threads. The cards were far easier to manufacture than the metal cylinders, and they could be arranged to create an infinite number of patterns. The automated nature of Jacquard's loom also made it more than twenty times faster than traditional drawlooms. "Using the Jacquard loom," James Essinger writes, "it was possible for a skilled weaver to produce two feet of stunningly beautiful decorated silk fabric every day compared with the one inch of fabric per day that was the best that could be managed with the drawloom."

The Jacquard loom, patented in 1804, stands today as one of the most significant innovations in the history of textile production. But its most important legacy lies in the world of computation. In 1839, Charles Babbage wrote a letter to an astronomer friend in Paris, inquiring about a portrait he had just encountered in London, a por-

Joseph-Marie Jacquard displaying his loom

trait that when viewed from across the room seemed to have been rendered in oil paints, but on closer inspection turned out to be woven entirely out of silk. The subject of the portrait was Joseph-Marie Jacquard himself. In his letter Babbage explained his interest in the legendary textile inventor:

> *You are aware that the system of cards which Jacard* [sic] *invented are the means by which we can communicate to a very ordinary loom orders to weave any pattern that may be desired. Availing myself of the same beautiful invention I have by similar means communicated to my Calculating Engine orders to calculate any formula however complicated. But I have also advanced one stage further and without making all the cards, I have communicated through the same means orders to follow certain laws in the use of those cards and thus the Calculating Engine can solve any equations, eliminate between any number of variables and perform the highest operations of analysis.*

Babbage borrowed a tool designed to weave colorful patterns of fabric, which was itself borrowed from a tool for generating patterns of musical notes, and put it to work doing a new kind of labor: mechanical calculation. When his collaborator Ada Lovelace famously observed that Babbage's analytical engine could be used not just for math but potentially for "composing elaborate . . . pieces of music," she was, knowingly or not, bringing Babbage's machine back to its roots, back to the "Instrument Which Plays by Itself" and Vaucanson's flute player. Always one to celebrate his influences, Babbage managed to acquire one of the silk portraits of Jacquard and displayed it prominently in his Marylebone home alongside Merlin's dancer and his Difference Engine.

The thread that connects Jacquard to Babbage to the digital pioneers of the 1940s is well-worn, for good reasons. Most histories of computation include a nod to Jacquard's punch cards, even though technically speaking he wasn't programming a computational device. But punch cards did persist as the dominant input and data storage device for digital information well into the second half of the twentieth century. (I can remember using punch cards as a grade-schooler in the 1970s.) Fittingly, punch cards were replaced as input devices by keyboards, and as storage devices by magnetic tape: both technologies, as we will see, that were originally designed to play or record music.

But there is another reason why tech historians are so quick to celebrate Jacquard's role in jump-starting the age of computation. The handoff from loom to analytic engine follows the wider pattern of economic paradigm shifts: the industry that launched the industrial revolution—textiles—provides the seeds for the digital revolution two centuries later. There's a kind of macroeconomic poetry to telling the story this way: one dominant mode of production laying the groundwork for one of its successors. But when you step back and look at the history from a wider angle—from the Banu Musa through the music-box curios to Vaucanson and his flute—you can't help noticing how long the idea of a programmable machine was kept in circulation by the propulsive force of delight, and not industrial ambition: first the patterns of sounds, produced by instruments that play themselves, then the patterns of color on a cloth. The entrepreneurs and industrialists may have turned the idea of programmability into big business, but it was the artists and the illusionists who brought the idea into the world in the first place.

In October of 1608, Ferdinando I, the Grand Duke of Tuscany and head of the Medici dynasty, hosted a monthlong pageant in Florence to celebrate the marriage of his son Cosimo to an Austrian archduchess. Given the vast wealth of the Medici clan, and the years of geopolitical negotiations that had preceded the engagement, the wedding was destined to be an extravagant affair: closer, in modern terms, to a city hosting the Olympics than a celebrity wedding. Jousting tournaments, equestrian ballets, even a mock naval battle on the Arno were all staged in the days leading up to the ceremony. The official wedding banquet was crowned with a theatrical production of a play (written by Michelangelo's son) interspersed with six musical numbers, known as *intermedi*, all based on new scores written for the wedding festivities by a team of court composers. The sixth and final *intermedio* featured an elaborate stage set in which the gods descended from the heavens to extend their welcome and congratulations to the newlyweds. A sketch of the stage design by young Michelangelo shows roughly a hundred musicians, singers, and dancers onstage—some of them apparently hovering above the audience in simulated clouds.

The tradition of lavish *intermedi*—which may have hit its climax with Cosimo's wedding banquet—is of particular interest to music historians, because it served as a kind of transitional form that would, over the course of the 1600s, solidify into the set of dramatic and orchestral conventions that we now call opera. But for our purposes, consider that extravagant performance purely in terms of the *technology* on display. Think first of the tools available to artists or scientists in the early years of the seventeenth century: the printing press was only a hundred and fifty years old; telescopes were just being deployed for the first time; microscopes were still fifty years away. Beyond the printing press, a writer's tools would have been indistinguishable from the quills and ink that the Greeks and Romans used more than two thousand years before. Painters had access to new oil paints that had been developed in the preceding century; some used the older technology of the *camera obscura* to create the distinctive realism of Renaissance painting. Needless to say, the technologies of photography and cinema were unimaginable to the creative class that the Medicis supported.

By contrast, the composers of the wedding banquet *intermedi* had at their disposal a rich and diverse array of musical technologies with which to entertain the Medicis and their guests. A roster of instruments that played during the final *intermedio*—an ensemble that bears a clear resemblance to a modern orchestra—gives a vivid picture of music's technological dominance in the early 1600s. To create their sonic extravaganza, the court composers drew upon multiple variations of plucked stringed instruments, including lutes, chitarroni, and citterns—instruments that date back at least to ancient Egypt. They employed a now largely extinct wind instrument called the cornett that was carved out of wood with a mouthpiece made of ivory, along with keyboard-based pipe organs that enabled long, sustained notes, an instrument that dates back to Roman times. Percussive sounds were generated by bronze cymbals and triangles. (Interestingly, the ensemble seems to have lacked any form of traditional drum.) The *intermedio* also featured metal trombones with a distinctive sliding tube to change pitch—a relatively new addition to the composer's palette, having been invented only a few centuries before. The most advanced instruments in the ensemble would have been the bowed string instruments: violins, violas, bass violins—technology so advanced that by most accounts it peaked less than a century after the Medici wedding with the production of the legendary Stradivarius violins still played by world-class soloists. And, of course, the ensemble included flutes, the descendants of those mammoth and vulture bone flutes from the Upper Paleolithic.

As diverse as the Florentine proto-orchestra might seem, several key instruments commonly used during that period were not featured in the performance: trumpets, harps, and, most important, the new keyboard-based clavichord and harpsichord. (The pianoforte, now called simply the piano, wouldn't be invented for another century or so.) Viewed purely in terms of the technological innovation at their disposal, the poets and painters of the early seventeenth century were living in the Stone Age compared to the high-tech bounty available to the composers. That extensive inventory of sonic tools illustrates

Legendary violin maker Stradivari's workshop

how, from the very beginning, our appetite for music has been satisfied by engineering and mechanical craft as much as by artistic inspiration. The innovations that music inspired turned out to unlock other doors in the adjacent possible, in fields seemingly unrelated to music, the way the "Instrument Which Plays by Itself" carved out a pathway that led to textile design and computer software. Seeking out new sounds led us to create new tools—which invariably suggested new uses for those tools.

Consider one of the most essential and commonly used inventions of the computer age: the QWERTY keyboard. Many of us today spend a significant portion our waking hours pressing keys with our fingertips to generate a sequence of symbols on a screen or page:

typing up numbers in a spreadsheet, writing e-mails, or tapping out texts on virtual keyboards displayed on smartphone screens. Anyone who works at a computer all day likely spends far more time interacting with keyboards than with more celebrated modern inventions like automobiles. Once the province of the great American novelists, clerks, and the typing pool, the keyboard has become so ubiquitous that we rarely celebrate it as an invention at all, preferring instead to pay tribute to the technically more advanced machines those keyboards are now connected to. But the idea of using the pads of our fingers to generate language—with individual keys mapped onto letters in the alphabet—was a tremendous breakthrough; without it, the digital revolution would have been at the very least significantly delayed. Technically speaking, keyboards were not as essential to advanced computation as the idea of programmability that Babbage inherited, indirectly, from the Banu Musa. But on a practical level, the existence of text keyboards allowed computers to achieve astonishingly complex feats. Other input mechanisms, like punch cards, were simply too slow. The chips might have followed the upward arc of Moore's law for a spell, but eventually the hardware would have hit a ceiling dictated by the software: without keyboards, writing code even for, say, a 1970s mainframe would have been tedious beyond belief. Voice entry might have one day offered an equivalently powerful medium for writing software, but it is unlikely that we would have been able to compose software smart enough to recognize spoken language without the speed and precision that keyboards supplied. You don't need keyboards to invent a computer, but it's awfully hard to get a computer do anything interesting without also inventing a typing mechanism to go along with it.

And yet, despite its obvious utility, even in the pre-digital era, the concept of an alphanumeric keyboard seems to have been strangely difficult for people to imagine. Inventors didn't begin tinkering with "writing machines" until the early 1800s, and the first commercially successful typewriter—the Remington No. 1—didn't go on the market until 1874. Like the bicycle, which wasn't invented until the middle of the 1800s, the typewriter is one of those technologies that seems to have emerged much later than it should have. Gutenberg had demonstrated the utility and commercial value of mechanized typesetting in the 1400s; Renaissance- and Enlightenment-era clockmakers—and automaton designers—had the engineering skills to design machines far more complex than the original Remington. "From a mechanical point of view," historian Michael Adler writes, "there is no reason why a writing machine could not have been built successfully in the fourteenth century, or even earlier." And yet no one bothered to do it.

What finally provoked inventors to start tinkering with typewriter keyboard designs? The answer is visible in the word itself: the "keys" we press when we type descend, etymologically, from the musical meaning of the word. While no one thought to build keyboards to trigger language until the nineteenth century, people had been constructing keyboards to make music since the pipe organs of Roman times. While Gutenberg was perfecting his movable type system, dozens of instrument designers across Europe were experimenting with different models of keyboard-based string instruments, devices that eventually solidified into those staples of Baroque music, the clavichord and harpsichord. In the early 1700s, the great-grandson of Cosimo II de' Medici—whose 1608 wedding had showcased such a wide range of musical devices—recruited an instrument maker named Bartolomeo Cristofori to manage his extensive collection of instruments. Cristofori went on to invent the pianoforte under the patronage of the Medici clan, a keyboard instrument that for the first time allowed notes to be played at different volume levels. (*Pianoforte* means "soft-loud" in Italian.) By the 1800s, the piano had become one of the most widely played instruments in the world.

Keyboard-based instruments were so appealing to musicians for the same reason that alphanumeric keyboards appeal to us today:

they enable us to use all ten of our fingers independently. And as the piano became increasingly prominent in the 1700s, inventors began to think about how the keyboard system might be used not just to play music but also to capture the notes played in some kind of permanent medium. In 1745, a German named Johann Freidrich Unger proposed a device that would draw musical notes on a rolling sheet of paper triggered by a live performance on a piano-style keyboard. Each note would be represented by a straight line, its length determined by the length of time the corresponding key was depressed. (Unger's proposed score looked remarkably like the MIDI "piano roll" layout used by digital music software today.) While the whole contraption depended on the paper roll unfurling at a constant speed, Unger was strangely silent about how his device would achieve that consistency, which may be one reason he appears to have never actually built the thing. Other inventors followed in his wake, including one Miles Berry, who patented a device in 1836 that punched holes through carbon paper with a stylus, a technique that would later become essential to player pianos.

But the flurry of interest in using keyboards to write musical notes suggested another application, with an even larger potential market: using keyboards to record letters, words, and whole sentences. Inspired by the way pianists played chords by depressing multiple keys simultaneously, a French librarian named Benoit Gonod invented in 1827 a shorthand typewriter using only twenty keys that printed a stream of dots that could be translated back into alphanumeric code. (Gonod's chordal system is still used by stenographers today.) The details are murky, but at some point in the 1830s or 1840s, an Italian named Bianchi appears to have presented to the Paris academy a "writing harpsichord" that used a piano keyboard and a cylindrical platen wrapped with paper to print actual letterforms directly on the page. The first functioning machine that a modern observer would identify as a typewriter was patented in 1855 by an Italian named Giuseppe Ravizza who had been obsessed with

Early Remington typewriter

the idea of writing machines for thirty years. He used the same metaphor that Bianchi had used before him, calling his creation the *cembalo scrivano*, the "writing harpsichord."

Other typewriter inventors during this period used piano-style keyboards—with interspersed white and black keys—including a "printing machine" invented by a New Yorker named Samuel Francis in 1857. But by the time the Remington No. 1 hit the market in the 1870s, the musical roots of the typewriter had been washed away, lingering only in the word *keyboard* itself. The typewriter keyboard was poised to reinvent the way humans communicate. But the *idea* for this now indispensable tool began in song.

Why did music play such an important role in our technological history? One likely reason is that music naturally lends itself toward the creation of codes, more than any other human activity other than language and mathematics. Cuneiform tablets dating back to 2000 BC have been found inscribed with a simple form of musical notation, featuring notes arranged according to what we now call the diatonic scale. Once again, music appears to leap ahead of where it should logically be on the hierarchy of needs. In 2000 BC, most human settlements around the world hadn't invented a notation system for *language* yet. And yet somehow the ancient Sumerians were already composing scores.

Rhythm, too, can function as a kind of informational code, as Samuel Morse discovered in the invention of the telegraph. The very first long-distance wireless networks were the "talking drums" of West Africa, percussive instruments that were tuned to mimic the pitch contours of African languages. Complex messages warning of impending invasions, or sharing news and gossip about deaths or marriage ceremonies, could be conveyed at close to the speed of sound across dozens of miles, through relays of drummers situated in each village. Instruments designed originally to set the cadence for dance and other musical rituals turned out to be surprisingly useful for encoding information as well. The origins of the talking drum technology are lost to history; there is no Samuel Morse to celebrate, some ingenious inventor of the original code. But presumably the idea followed roughly the same sequence that brought other civilizations from bone flutes to music boxes: the patterns of synchronized sound were initially triggered for the strange intoxication they invoked in their listeners. But, over time, West African inventors, like Vaucanson so many centuries later, began to notice that those patterns could convey something more than mere rhythm or melody. Like the patterns of wave forms that we translate into spoken language, the patterns of tones generated by the drums could carry another layer of meaning. For the first time, symbols were aloft in the air, traveling far beyond the range of the human voice.

Today, of course, our lives are surrounded by encoded information and entertainment. We watch movies and share family photographs and play games that have been translated and compressed into binary code, captured in some kind of storage medium, distributed through optical disks or, increasingly, through the Internet—and then, at the very end of the chain, the code is translated back into sounds and images and words that we can understand, by machines designed to translate these hidden codes for us, turning them into meaningful information. This vast cycle of encoding and decoding is now as ubiquitous as electricity in our lives, and yet, like electricity, the cycle was for all practical purposes nonexistent just a hundred and fifty years ago.

Not surprisingly, one of the very first technologies that introduced the coding/decoding cycle to everyday life took the form of a musical instrument, one with a direct lineage to Vaucanson and the Banu Musa: the player piano. Though its prehistory dates back to the House of Wisdom, a self-playing piano became a central focus for instrument designers in the second half of the nineteenth century; dozens of inventors from the United States and Europe contributed partial solutions to the problem of designing a machine that could mimic the feel of a human pianist. The new opportunities for expression that Cristofori's pianoforte had introduced posed a critical challenge for automating that expression; it wasn't enough to record the correct sequence of notes—the player piano also had to capture the loudness of each individual note, what digital music software now calls "velocity." The first player piano to reach genuine commercial success was the pianola, designed by a Detroit inventor named Edwin Scott Votey in 1895. Powered pneumatically by the suction created

When Summertime Ends

When Summertime ends, and thoughts turn to the cosiness of drawn blinds and fireside recreations, is the time to purchase your 'Pianola' Piano.

For you the 'Pianola' Piano will fill the winter evenings with the joy of music making – all the music there is, grave or gay, sonata, foxtrot, song accompaniment, the wonderful 'Pianola' controls are so simple to operate, yet so sensitive, that with very little practice you can play all classes of music with perfect expression.

Newspaper advertisement for the Pianola, 1920s

by a human pressing foot pedals (which in turn determined the tempo of the performance), the Pianola encoded its songs onto perforated rolls of paper—a scrolling version of Jacquard's punch cards. While dozens of other player pianos swarmed the market in the early 1900s (including one called the Tonkunst—German for "musical art"—from which the phrase *honky-tonk* derives), the pianola was such a success that it achieved Kleenex and Band-Aid levels of brand awareness. During the heyday of the player piano—which lasted until around 1930—many consumers simply called them pianolas regardless of the device's actual brand name.

With the commercial success of the pianola came an innovative new business model: selling new songs that had been encoded in the piano roll format. Today we take this exchange for granted: you buy a piece of hardware and then spend the next few years purchasing software to run on it, code that endows the machine with new functionality. But in 1900, the whole concept of paying for new programming was an entirely novel idea. Hundreds of thousands of songs—from classic compositions to honky-tonk—were recorded in piano-roll format in the first decades of the twentieth century. And while the player piano would be largely killed off by radio and phonographs by the 1930s, the commercial template it established—paying for code—would eventually give rise to the some of the most profitable companies in the history of capitalism.

There is something puzzling about the fact that the player piano achieved such dramatic success during this period. Radio, of course, didn't become a mainstream technology until the 1920s, so the fact that the pianola found a foothold in American homes and public venues before then shouldn't surprise us. But the phonograph was invented by Edison in 1877, almost twenty years *before* Votey's pianola. A phonograph was far less cumbersome as a device—and less expensive—and it would go on to replace the player piano as a "programmable" machine by the 1930s. So why did it take so long? As is so often the case in history of innovation, the explanation for this strange sequence revolves around the networked nature of technological progress. Inventions are almost never solitary, isolated creatures; they depend on other inventions that complete them, or endow them with new applications that their original inventors never considered. The phonograph on its own was a true breakthrough, capturing the actual analog wave forms of music and human speech for the first time. But it required another, parallel technology to reach mainstream success. A pianola was at its core an actual piano, with hammers hitting actual strings; it generated sound loud enough to fill a clamorous parlor or saloon. But the phonograph was amplified only by the passage of the sound waves through a flaring horn. You had to lean in to hear it, and in a room with any sort of ambient noise, the sound was effectively inaudible. The phonograph was far more versatile than the pianola; you could hear singers, brass bands, orchestras, or poetry. It just wasn't loud enough. It needed amplification, which arrived via two related inventions: metered electric currents and vacuum tubes. The second you could plug in the phonograph, the pianola's days were numbered.

The distinguished crowd gathered on June 19, 1926, at the Théâtre des Champs-Élysées, the grand art deco concert hall in the eighth arrondissement, would have known they weren't in for just another night at the opera the second they caught sight of the instruments assembled on the stage. Alongside a handful of percussive instruments that wouldn't have been out of place at the Medici wedding—xylophones, glockenspiels, bass drums—the ensemble also included a pianola, several traditional pianos, a siren, hammers, saws, a collection of electric bells, and two oversized airplane propellers.

The musicians playing these bizarre instruments were there to perform a piece called *Ballet Mécanique*, written by the twenty-four-year-old American composer George Antheil. Like many Jazz Age artists and intellectuals, Antheil had left the banal Americana of his upbringing and set sail for the bright lights and bonhomie of the Parisian avant-garde at the age of twenty-one. By the time of the Théâtre des Champs-Élysées premiere, he'd made a home for himself among the cognoscenti: renting a room above Shakespeare and Company from Sylvia Beach; dreaming up collaborations with James Joyce and Ezra Pound; feuding with Igor Stravinsky. Drawing on the Futurists' obsession with the aesthetics of industrial machinery, Antheil built a reputation as a daring composer eager to widen the definition of what constituted a musical instrument in the first place. (Years later, Antheil, in an act of self-branding that would have impressed Prince, titled his autobiography *The Bad Boy of Music*, although some of his musical innovations mimicked earlier experiments by Stravinsky and Darius Milhaud.) *Ballet Mécanique* had originally been conceived as a score for a short experimental film by the same title, but the two works soon took on separate lives—in part because Antheil's score turned out to be almost twice as long as the film.

As eclectic as the "orchestra" might have seemed on the stage of the Théâtre des Champs-Élysées, it represented a far more streamlined and conventional ensemble than Antheil had envisioned. The centerpiece of his original score was a suite of sixteen pianolas playing four distinct parts, many of them atonal. The "mechanical ballet" that Antheil initially planned to stage had automated machines instead of dancers—the pulsing rhythm and noise of the industrial age transformed into a kind of music, not through the dexterity of the virtuoso player, but through the punched notes of the piano roll. And yet, ultimately, the pianola technology undermined Antheil's project, in large part because he was asking the technology to do something it was almost never asked to do in ordinary use. A piano roll could effectively mimic the tempo and articulation of an accomplished pianist, coordinating the activity of all eighty-eight notes on the keyboard. But there was no easy way to synchronize the activity of *multiple* player pianos. Even if you triggered each of them at the exact same time, minuscule differences between the tempos of each machine would quickly cause them to fall out of sync. Today, modern concerts effortlessly sync dozens of different instruments thanks to the digital code of MIDI, but in the midtwenties this was a problem that no one had even conceptualized yet. Instruments were locked to the same tempo because they were being played by human beings who could follow the gestures of the conductor or hear the beat being generated by the rhythm section. But a machine needed mechanical cues to stay in sync with other machines. Antheil had attempted to jerry-rig a system of pneumatic tubes and electric cabling to coordinate the pianolas, but none of his efforts worked in the end. Eventually, he gave up and rewrote the score for a single pianola, accompanied by traditional pianists.

Even without the sixteen player pianos, the 1926 premiere managed to provoke a small riot in the theater. With an all-star lineup of Parisian intellectuals in attendance—including Joyce, Pound, T. S. Eliot, and Man Ray—the crowd began to hurl invectives at the stage as the performance rumbled on; when the airplane propellers lurched into action, concertgoers opened up umbrellas to pantomime being blown away by the airflow. A friend of Antheil's later wrote, "The Ballet began to seem like some monstrous abstract beast, battling with the nerves of the audience, and I began to wonder which one would win out." At one point, Pound apparently stood up and shouted to the catcallers: *"Vous êtes tous des imbéciles!"* ("You are all imbeciles!"). Aaron Copland later suggested that the *Ballet Mécanique* riots were even more tumultuous than the famous outburst that greeted the premiere of Stravinsky's *Sacre du Printemps.* (In Copland's words, Antheil had "outsacked the Sacre.") Antheil, of course, saw the whole affair as a great triumph: "From this moment on," he wrote in his autobiography, "I knew that, for a time at least, I would be the new darling of Paris."

The *Ballet Mécanique* performance has a kind of quaint nostalgia to it now: those were the good old days, when experimental music performances in concert halls used to provoke mayhem in the aisles.

But I think that night in 1926 embodies a set of deeper truths about the way music has long been intertwined with technological innovation. Something about the nature of music encourages subcultures to seek out new kinds of sound, often by designing new instruments, like the synchronized player pianos of Antheil's original score. Eventually, those new sounds—as harsh or unintelligible as they might be to the first generation of listeners—grow more familiar; mainstream composers and performers integrate them into more traditional songs and arrangements, the way electronica and dubstep have become a cornerstone of much of today's Top 40. And that exploratory urge for new sounds ends up creating new technological possibilities that have applications outside the realm of music. Antheil failed to devise a scheme to synchronize his sixteen player pianos. But the attempt to solve that problem—in the name of art and sonic experimentation—would lead, almost two decades later, to a breakthrough military technology, one that would eventually become a crucial component of civilian wireless communications. As it happens, to make that leap, Antheil needed the most unlikely partner imaginable: Hedy Lamarr, at the time one of the most glamorous movie stars in the world.

Ignored for many decades, the story of Lamarr and Antheil's strange-bedfellows collaboration has in recent years become part of tech-history lore. Lamarr had begun her career as an actress in Weimar Germany, and married an arms dealer who ultimately turned out to have ties to the Nazis. She fled Europe in 1937 and found her way to MGM studios, where she quickly became one the great seductresses of American cinema. Offscreen, though, Lamarr was an amateur inventor who was constantly tinkering with ideas that might help the Allied cause. She largely avoided the Hollywood party circuit, preferring to stay at home in a living room populated by engineering textbooks and drawing boards pinned with schematics. After a German U-boat sank the refugee ship *City of Benares* in September of 1940—killing seventy children—Lamarr began sketching out a

Hedy Lamarr

plan for a remote-controlled torpedo that could take out a submarine. She was well aware, from her days on the periphery of the Austrian munitions business, of the difficulty with wireless control mechanisms. Using radio frequencies, it was easy enough to send a signal that could alter the direction of a torpedo in the water; the problem was that those signals were easily detected and jammed by the enemy. But it occurred to Lamarr that a system could route around that interference by constantly changing frequencies according to some preestablished pattern. She called this technique "frequency hopping."

Fortuitously, Lamarr happened to meet Antheil just as this idea was taking shape in her mind. Since his riotous days performing *Ballet Méchanique*, Antheil had followed a Fitzgeraldian path from Paris to Hollywood, where he developed a successful, though less scandalous, career composing scores for a series of forgettable films. (Amazingly, he also wrote a relationship advice column for *Esquire*.) After dining together several times in the fall of 1940, Lamarr and Antheil began working on the frequency-hopping project as a team. It may well have been the most bizarre partnership in the history of invention: the movie star and the experimental composer/advice columnist, working late into the night in the Hollywood Hills, brainstorming cutting-edge ideas for naval communications protocols.

Lamarr's original frequency-hopping idea, as brilliant as it was, presented one critical challenge: the receiving mechanism had to shift frequencies in concert with the sending device. It was, in a sense, a synchronization problem. This is where Antheil's experience with *Ballet Mécanique* supplied the missing element that completed Lamarr's invention. Sitting in Lamarr's drawing room night after night, wrestling with the problem of coordinating a shifting set of frequencies in perfect tempo, he found himself thinking back to those sixteen pianolas and his dream of programming them all in perfect time using barrel rolls. He proposed a control system whereby the instructions for frequencies were encoded in two perforated ribbons. Where the holes in the piano roll signaled a musical note, the holes in the ribbons signaled a frequency change. To an enemy scanning the spectrum for signals to jam, the pattern would be impossible to detect. But as long as the ribbons were activated at the same time, to the receiver on the torpedo the random bursts of noise jumping across the spectrum would sound like information. As a nod to the musical roots of the idea, Antheil proposed that the system include eighty-eight distinct frequency hops, the exact number of keys on a piano.

While the navy largely ignored Lamarr and Antheil's proposal, the two inventors did receive a patent for it in 1942, and it was considered valuable enough to be locked away as classified information. (Antheil later wondered if the musical analogy hurt their cause with the hardheaded military analysts: "'My god,' I can see them saying, 'we shall put a player piano in a torpedo,'" he wrote.) But the idea of frequency hopping was too good to disappear forever in the vaults of classified patents. An updated version of Antheil and Lamarr's invention was implemented on navy ships during the Cuban missile crisis. Today, frequency hopping has evolved into the spread-spectrum technology used by numerous essential wireless systems, including cell-phone networks, Bluetooth, and Wi-Fi.

George Antheil correctly foresaw the extraordinary explosion of new sounds that the age of electricity promised, though he may have been a bit shortsighted about the data storage technology those sounds would employ. "We shall see orchestral machines with a thousand new sounds, with thousands of new euphonies, as opposed to the present day's simple sounds of strings, brass, and woodwinds," he wrote in his 1924 "Mechanical-Musical Manifesto." "It is only a short step until all [musical performance] can be perforated onto a roll of paper." Reasonable people can disagree about whether strings or brass or woodwinds are "simpler" than the new sounds unleashed by electronic music, but it is an empirical fact that "orchestral machines" in the form of synthesizers, samplers, and digital music software have introduced thousands of new sounds to the modern listener. And these novel sonic textures are not just being savored by the educated ear of the avant-garde aficionado; they are ubiquitous on Spotify and SiriusXM. No doubt there would have been even more outrage if that 1926 audience had been exposed to Kraftwerk or Aphex Twin. But today those machine-made—and often noise-saturated— sounds have found their way into mainstream taste.

Like almost every innovator in the history of instrument design, the pioneers of electronic music charted a path that would eventually be followed by other, less adventurous settlers. Today we take photographs and shoot films with digital cameras; we write novels on digital computers; we design buildings using CAD software. But the first electronic and digital tools for creativity were almost exclusively musical in nature. Some of those tools—and the people behind them—became recognizable names, like the Moog synthesizer designed by the brilliant engineer Robert Moog. (Always willing to experiment with new sounds, the Beatles were among the first musicians to integrate the Moog into the recordings—you can hear it playing the organlike parts on "Here Comes the Sun.") But as with any paradigm shift worth the title, brilliant mistakes or generative dead ends propelled the change almost as forcefully as the success stories.

One of those dead ends resides in the collection of London's Science Museum. With coils of cables and film ribbons streaming out of an open frame the size of a chest of drawers, it brings to mind an office photocopier that's been dissected on an autopsy table. But this strange device is, in fact, a musical instrument, one of the very first functioning electronic instruments ever built. It is called the Oramics Machine, named after its creator, Daphne Oram.

Oram belonged to a generation of women who played crucial roles in designing and programming some of the first machines of the digital age. Oram herself was a sound engineer at the BBC, and a composer of experimental music. (Taking up where Antheil left off, she tinkered with compositions using multiple tape recorders and sine wave oscillators.) One of her first jobs at the BBC happened to coincide with the waning days of the Blitz; during live radio broadcasts of classical performances, Oram would synchronize recorded performances of the music. If a German bomb strike interrupted the performance, she would deftly switch over to the recorded version. (Given the poor audio quality of AM radio and 1940s-era speakers, the listening audience at home rarely noticed the handoff.) By the 1950s, she had built up enough political capital inside the BBC to successfully argue for the creation of the BBC Radiophonics Workshop, an influential sound-effects lab that endured for forty years.

The idea for the Oramics Machine had its origin during her technical training course in 1944. For the first time, Oram encountered a cathode-ray oscilloscope, a device that converts incoming sound waves into a scrolling, jittery line on a screen, not unlike a modern EKG. Oram recalled the experience decades later:

> *I was allowed to sing into it and there I saw my own voice as patterns on the screen, graphs, and I asked the instructors why we couldn't do it the other way around and draw the graphs and get the sound out of it. I was eighteen I think and they thought this was a pretty stupid, silly teenage girl asking silly questions. But I was quite determined from that time on that I would investigate that, but I had no oscilloscope.*

Oram's question may have baffled her instructors, but something about the idea of drawing sound stuck in the back of her mind. When the world of electronics matured in the 1950s, with widespread adoption of transistors, photocells, and magnetic tape, Oram became convinced that the time was right to turn her teenage hunch into a reality. She first went to the BBC, arguing that they should fund the development of a machine that could, as she later put it, "[explore] this vast new continent of music." The top brass at the BBC found the request baffling. "I went to see the Head of Research and said I've got an idea of writing graphic music could I have some equipment please," Oram recalled. "He pulled himself up to his full height and said 'Miss Oram, we employ a hundred musicians to make all the songs we want, thank you.' . . . That was the official attitude: they had the BBC Symphony Orchestra, and it was there to make all the music they wanted, and nothing else was of any interest."

Rebuffed by her lifelong employer, Oram struck out on her own in 1962 and began working on her Oramics device independently, supported by a series of grants and a rotating cast of engineers. Reminiscent of the perforated ribbons that Antheil and Lamarr had devised, the device used strips of 35mm film to control the sound. The "composer" would sketch a wavering line that designated changes in timbre, amplitude, or pitch. Photocells scanned the ribbons and passed the information on the sound generators. The final iteration of the machine included ten separate film strips that could be simultaneously processed. The device had four "voices" that could play distinct parts in perfect tempo. Notes were programmed with a grid of rectangles that took its cues from the piano-roll format. The other elements—volume, say, or reverb—were controlled by free-form shapes sketched onto the film. The Banu Musa and Vaucanson had struggled with the difficulty of "cutting" a pinned cylinder to program a musical machine. Daphne Oram thought it should be as simple as drawing a line.

In a 1966 letter, she described her progress:

> *We are delighted to tell you that we have succeeded in proving that graphic information can be converted into sound. We can draw any wave form pattern and scan this electronically to produce sound. By varying the shape of the scanned pattern the timbre is varied accordingly . . . We believe that no similar piece of equipment exists anywhere else in the world.*

Daphne Oram drawing timbres on the Oramics machine

She was correct about that belief. Not only was the Oramics the first of its kind, it was, in a way, the last of its kind. Oram had been right to argue that a "vast new continent of music" was about to surface. By the middle of the 1970s, adventurous musicians were recording entire albums of electronic tones—but the synthesizers they used relied on much more traditional input mechanisms: the tones controlled by a dashboard of knobs and dials, and notes controlled by that most ancient of conventions—the keyboard. By the time electronic music became mainstream thanks to postpunk bands like Devo and New Order, controlling a synthesizer by painting lines on a ribbon of film would have seemed as pointless as controlling it with punch cards.

But we shouldn't be distracted by the spectacle of those film strips. Yes, that particular part of the solution turned out to be a dead end. Computer interfaces, not undulating lines painted onto film strips, turned out to be much more efficient at the task of telling the synthesizers what notes to play. But look at any modern piece of music recording software—Pro Tools, Logic, even GarageBand; the default view of all of these now-indispensable tools is a stack of undulating lines, scrolling across a screen in time with the song. Some of those lines control pitch; some control volume and reverb. There were hundreds of provocative electronic instruments under development during the sixties, many of which produced more direct descendants than Oram's machine. But none of those devices have the striking family resemblance that modern software tools share with the Oramics machine. The physical medium Oram devised was obsolete almost as soon as it was invented. But the *interface* she devised belonged to the future.

The technological sequence that musical instruments followed—from mechanics to electronics to software, from the physical notches in the piano roll to the invisible zeros and ones of code—also shaped the evolution of *recorded* music: the windup gramophone replaced by the tube amps of stereo gear replaced by the digital bits of audio CDs. The migration to digital music is recent enough that most of us grasp its wider repercussions: think of the still-simmering debates over music in the post-Napster era, the challenges it has posed to our intellectual property laws and the economics of all creative industries. But, as always, what began with an attempt to create and share new kinds of sounds ended up triggering other revolutions in other domains. The first true peer-to-peer networks for sharing information were designed specifically for the swapping of musical files. It is still too early to tell, but this innovation may turn out to be as influential as those piano keyboards and pinned cylinders, if in fact peer-to-peer platforms like Bitcoin eventually become an important part of the global financial infrastructure, as many people believe. It is entirely possible that the most significant advance in the history of money since the invention of a government-backed currency will end up having its roots in teenagers sharing Metallica songs.

Whenever waves of new information technology have crested, music has been there to greet them. Music was among the first activities to be encoded, the first to be automated, the first to be programmed, the first to be digitized as a commercial product, the first to be distributed via peer-to-peer networks. There is something undeniably pleasing about that litany, something hopeful. Think about the history behind the most influential device of the modern age: a digital computer sharing information across wireless networks. What were the enabling technologies that made it possible to invent such a device in the first place? The standard story is that computers—and the Internet—descend from military technology, since many early computers were designed specifically to crack wartime codes or calculate rocket trajectories. But inventing a computer also required other building blocks: music boxes, automated flautists, harpsichord keyboards, player pianos. Too often we hear the old bromide that innovation invariably follows the lead of the warriors. But it turns out the minstrels and the maestros led us to their fair share of breakthroughs as well, particularly with technology that involves some kind of code. Yes, the Department of Defense helped build the Internet. But the pinned cylinders of the music boxes gave us *software.* When it comes to generating new tools for sharing and processing information, the instruments of destruction have nothing on the instruments of song.

3

Taste

The Pepper Wreck

Deep history can usually be detected in the most banal of artifacts, if you know where to look. Consider one of the most derided items on a modern supermarket shelf: the humble Doritos chip. Originally introduced in 1964 as an exclusive treat for Disneyland visitors, showcased at the "Casa de Fritos" in Adventureland, Doritos have become the crown jewel in the Frito-Lay empire of snacks. Since that amusement-park debut, the Doritos family has diversified into a wide range of flavors and cobranding opportunities: Cool Ranch, Sour Cream and Onion, Taco Bell's Taco Supreme, Pizza Hut's Pizza Cravers, Nacho Chipotle Ranch Ripple, Four Cheese, Spicy Nacho. Like every food sold in America since 1990's Nutrition Labeling and Education Act, the packaging of Doritos contains, in near-microscopic print, a list of all the ingredients used to produce the chips:

> *Whole corn, vegetable oil (corn, soybean, and/or sunflower oil), salt, cheddar cheese (milk, cheese cultures, salt, enzymes), maltodextrin, whey, monosodium glutamate, buttermilk solids, Romano*

cheese (part skim cow's milk, cheese cultures, salt, enzymes), whey protein concentrate, onion powder, partially hydrogenated soybean and cottonseed oil, corn flour, disodium phosphate, lactose, natural and artificial flavor, dextrose, tomato powder, spices, lactic acid, artificial color (including Yellow 6, Yellow 5, Red 40), citric acid, sugar, garlic powder, red and green bell pepper powder, sodium caseinate, disodium inosinate, disodium guanylate, nonfat milk solids, whey protein isolate, corn syrup solids.

All those ingredients, assembled to produce a two-dollar bag of snack food. Put aside all the chemistry-lab mystery ingredients we associate with processed junk food—sodium caseinate, disodium inosinate, disodium guanylate—and consider just the ingredients you actually recognize as food. Every Doritos chip offers a reminder of how globally intertwined our food networks have become. We think of Frito-Lay products as the ultimate highway convenience-store nonfood, but in a way, they are true citizens of the world. Corn was originally domesticated as maize in Mexico; soybeans first took root as an ancient East Asian crop; sunflowers were mostly native to North America; cheddar cheese was first crafted in England, while Romano comes from Italy. The milk in buttermilk and other cheeses dates back to the first cows that were domesticated for milk in Southwest Asia ten thousand years ago. No one knows for sure where onions first originated, but they are likely as old as agriculture itself. While we think of tomatoes as staples of the cuisines of Spain and Italy, the tomato plant first grew in the Andes of South America. Sugarcane hails from Southeast Asia, garlic came from Central Asia, and red and green pepper were native to Central and South America.

An entire planet's worth of flavors converge every time you savor the tangy, sharp taste of that Doritos chip. How did this globalized palate first come into being? The answer to that question is right there on the Doritos' packaging, in the most enigmatic item on the ingredients list: spices.

Roughly three hundred kilometers east of the Indonesian mainland lies a string of small, tropical islands, verdant volcanic cones that formed only ten million years ago. Today, geologists understand that they reside at a unique position on the planet's surface: the only place on Earth where four distinct tectonic plates converge. Formally, they are called the Maluku archipelago, but the five islands at the northern tip of the archipelago—Ternate, Tidore, Moti, Makian, and Bacan—have long been known by another name: the Spice Islands. Until the late 1700s, every single clove consumed anywhere in the world began its life in the volcanic soil of those five islands.

Despite their remote location and diminutive size, the islands have served as nodes on a global network of trade for at least four thousand years. Archeologists in Syria have found cloves preserved in ceramic pottery at the Old Babylonian site of Terqua, dating back to 1721 BCE. We think of the spice trade as a practice that belongs to the Age of Exploration, but its roots are far older. Somehow, in an era before compasses, accurate cartography, or printing presses, word had spread across the planet of the clove's alluring taste and aroma, and a network of trade had assembled to transport these tiny flower buds six thousand miles, from the Molucca Sea to the banks of the Euphrates.

The sheer scale of the transportation network that brought the cloves to Syria seems almost impossible given the navigational limitations of the age. A Babylonian living two thousand years before Christ would have had no idea about the existence of the Spice Islands, or Indonesia—or the Indian Ocean for that matter. Put another way, those tiny spices traveled thousands of miles farther than any individual human had ever traveled. The epic relay race that brought the cloves to modern-day Syria is lost to us now, but our knowledge of spice-trade routes from Roman times suggests the broad outlines of the itinerary. Outriggers or Chinese traders would

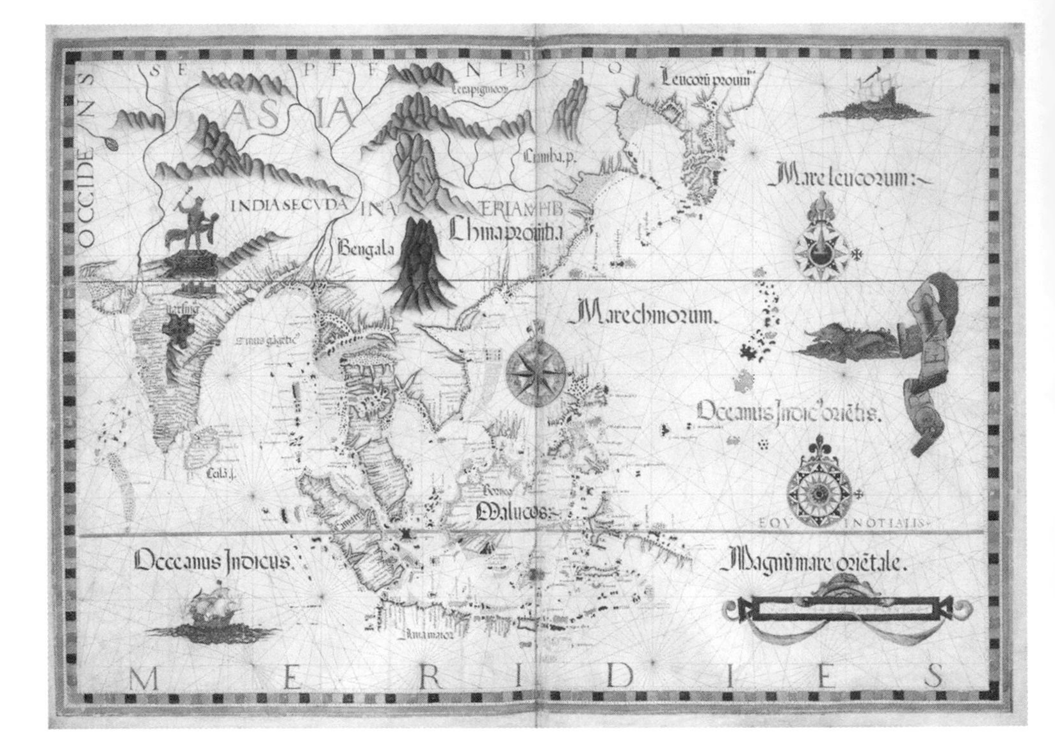

Map of the Spice Islands

have carried the cloves down through the Java Sea, winding their way through the narrow straits of Malacca. At port cities in Sumatra or modern-day Malaysia, they would have been sold to Indian traders who brought them across the Bay of Bengal around Sri Lanka to the Malabar coast of India. There Arab ships would have transported them up through the Persian Gulf, off-loading them to desert caravans whose camels would pull the cloves across modern-day Iraq to the kitchens of Babylon.

The first group to replace that distributed network of local traders with a single integrated global system were the Muslim traders who came to prominence in the seventh century CE. The new regime did away with the regional intermediaries, creating a massive and unified market that stretched from the Indonesian archipelago to Turkey and the Balkans all the way across sub-Saharan Africa, following the Senegal and Niger rivers. Muslim traders worked the entire length of this vast system, and their interactions with local communities introduced more than just a taste for cloves; they also brought Islam to these regions of the world. In almost all the places where Muslims attempted to convert local communities through military force—Spain or India, for instance—the Islamic faith failed to take root. But the traders turned out to be much more effective emissaries for their religion. The modern world continues to be shaped by those conversions more than a millennium later. The map of the Muslim spice trade circa 900 CE corresponds almost exactly to the map of Islamic populations around the world today. Indeed, it is entirely possible that Islam would not have become a major global religion without the long reach of the spice traders' integrated network. The geography of Islam in the twenty-first century is, in effect, the afterimage of a much earlier map: places where Muslims turned a profit introducing delightful new flavors to the taste buds of consumers.

The clove was not the only botanical jewel produced by the islands east of Indonesia. South of the Malukus, an even smaller archipelago known as the Banda Islands, with a combined landmass of only seventeen square miles, was the exclusive home of the nutmeg tree until modern times. The archipelago included two islands, Pulau Ai and Pulau Run, historically anglicized as Puloway and Puloroon. The latter island is so small one would be hard-pressed to fit a regional airport on its soil, and yet, at the height of the spice trade, James I formally referred to himself as "King of England, Scotland, Ireland, France, Puloway and Puloroon." A few decades later, the British would surrender their holdings in the Bandas to the Dutch, in return for a slightly larger, but less botanically unique, island on the other side of the world: Manhattan.

When Columbus returned triumphantly in 1493 to announce to

Engraving of nutmeg, circa 1798

the Spanish court that he had succeeded in finding a westward passage to the Orient, he brought back three main pieces of evidence to support his claim: parrots, a small retinue of Caribbean natives that he mistakenly called Indians—and cinnamon. (In the end, he was as wrong about the cinnamon as he was about the Indians; the spice turned out to be the bark of an unrelated Caribbean tree and tasted almost nothing like cinnamon.) Risking one's life and vast amounts of money to sail across an uncharted globe—all in pursuit of condiments? It seems almost comical to us now. But while the taste for spice may seem frivolous to us today, that desire—*appetite* is the wrong word for foodstuffs with such limited nutritional value—is the origin point for the cosmopolitan mélange of flavors baked into that Doritos chip, not to mention the international melting pot that is every modern supermarket. More than that, the seeds and buds of cloves, cinnamon, and nutmeg were also the seedlings out of which the whole idea of a global marketplace first sprouted. As the historian Jack Turner writes in his authoritative account, *Spice: The History of a Temptation*, "Nowhere is the history of East and West more incestuous than at the table. For the sake of spices East and West had an ancient relationship. In light of the appearance of spices in the most remote periods, it is a reasonable possibility that it was because of spices that they first met." And no spice did more to transform the planet than the one that now graces every dining room table, so ubiquitous that it is often distributed for free: pepper.

In September of 1606, a Portuguese cargo ship named the *Nossa Senhora dos Martires* arrived at the mouth of the Tagus River, not far from Lisbon. The ship was returning from a nine-month voyage, its hulls filled with a small fortune of goods brought back from India. But before the *Nossa Senhora dos Martires* could make her way to the Lisbon anchorage through the dangerous northern channel of the Tagus, the wind fell and she was dragged against the rocks of the São

Julião da Barra promontory. The intensity of the collision and the heavy winds caused the structure of the ship to fail catastrophically. The ship settled into the riverbed in pieces near the newly constructed fortress of São Julião da Barra, where she remained for almost four centuries.

As the years passed, the legend of a sunken treasure in the shadow of São Julião da Barra grew. Scuba divers began exploring the area in the 1950s and caught glimpses of a ship's hull buried in the silt. In 1996, the Portuguese Museum of Archeology sponsored an excavation of a one-hundred-square-meter site where a large section of the hull had settled. The promise of a vast hoard of treasure turned out to be myth: the archaeologists recovered pieces of Burmese stoneware, porcelain, and a handful of gold and silver coins. But the real treasure of the *Nossa Senhora dos Martires* was plain to see: the coins and trinkets and pottery lay buried beneath a dark blanket of peppercorns.

Worthless to a modern treasure hunter, the peppercorns would have been worth millions had the *Nossa Senhora dos Martires* made it to safe harbor in Lisbon. When the ship sank back in 1606, it left behind a black tide of peppercorns that washed ashore on the banks of the Tagus, where they were eagerly scavenged by local residents who flocked to the riverside to collect the spice. Today, the "Pepper Wreck"—as it has come to be called—is a startling reminder of just how valuable this now-quotidian spice was during its prime. During the Middle Ages, a pound of pepper was at many points worth more than a pound of gold. (Today a pound of gold will set you back almost $20,000, while a pound of pepper can be acquired for around five dollars.) Peppercorns were regularly used as a form of payment, serving as a kind of "universal currency," as Turner calls them. The custom of paying (or supplementing) one's rent with a few pounds of pepper continued in parts of Europe until the 1900s. When the Portuguese queen Isabella married in 1526, a significant portion of her dowry came in the form of peppercorns.

Divers raise a section of the "Pepper Wreck"'s keel

No civilization in history went as insane over spices as the Roman Empire. An average meal served at a banquet for the Roman elites would include an entire spice rack's worth of flavoring. Apicus's *Cookbook* from the first century CE features pepper prominently in 80 percent of its recipes, including "a variety of spiced desserts such as a peppered wheat-flour fritter with honey or a confection of dates, almonds, and pine nuts baked with honey and a little pepper." Fittingly, when the Barbarians laid siege to Rome in 408 CE, their leader, Alaric the Goth, offered to call off the blockade in return for a bounty of gold, silver, silk, and three thousand pounds of pepper. That may sound absurd—like asking for a bounty of plastic forks and paper napkins—but the Roman taste for spice had momentous consequences. Some historians believe that the Roman trade deficit with India—consisting in large part of pepper imports—played a critical role in triggering the fall of the empire itself. In his *Natural History*, Pliny the Elder noted, "There is no year in which India does not drain the Roman Empire of fifty million sesterces." The legendary decadence of the late Roman Empire wasn't just about moral decline; those spice-laden feasts had economic costs as well.

If pepper helped trigger the collapse of one of Europe's great cities, it helped build others. A modern visitor to the canals of Venice or Amsterdam, admiring the palazzos along the Grand Canal or the elegant town houses ringing the Herengracht, would do well to pause for a moment and consider that much of this worldly sophistication was originally funded by spices. Venice became the central European distribution point for pepper and other spices in the mid-thirteenth century, after Muslim traders had brought the spices to the Adriatic from India in caravans. The profits Venice made from the sale of its legendary Murano glass were an afterthought compared to the tariffs it charged as a middleman in the spice trade. By the 1600s, the direct sea routes to India had already started to lower the price of pepper on the open market, though it was still precious enough to inspire a new age of empires and global corporations. The

Dutch East India Company—the very first corporation to issue publicly traded stock—was founded to exploit the immense profitability of the spice trade. Modern economists estimate that the Dutch markup on nutmeg and clove was as much as 2,000 percent.

The taste for pepper and other spices also provoked some of the darkest chapters in the history of globalization. The British East India Company instituted a virtual slave colony on the island of Sumatra to increase the productivity of its pepper vines there. In the late 1600s, the Dutch began trading opium grown in India in return for Indonesia pepper, triggering a plague of addiction that would last for centuries. An English trader wrote in 1711, "The Mallayans [now Malaysians] are such admirers of opium that they would mortgage all they hold most valuable to procure it." A century later, American traders would embrace the same brutal tactics. "Nothing is more certain than that opium brings generally 100 percent [profit] when sold to the Malays in Barter," an American merchant named Thomas Patrickson wrote to a friend in 1789, "and no reason can be alleged against visiting the Malay Coasts except Danger."

In terms of sheer brutality, few regimes in the history of the spice trade rival the Dutch domination of the Spice Islands that began in the early 1600s. When the Bandanese population had the audacity to challenge the suggestion that a handful of Europeans should have exclusive rights to the islands' bounty, the Europeans went into a fury of mercantilist genocide that was breathtaking in its speed and efficiency. In what the historian Vincent Loth called "one of the blackest pages in the history of Dutch overseas expansion," the Dutch, led by Governor-General Jan Pieterszoon Coen, managed to kill thirteen thousand Bandanese natives of the island Lonthor in a few weeks of organized savagery. Coen brought in Japanese mercenaries to assassinate the Bandanese elite, forty of whom were decapitated, their heads displayed on spikes to stifle further dissent. In a matter of decades, the indigenous population of the Bandas had vanished from the islands, most of them murdered. In their place, the

The first Dutch ship in the East Indies, 1596 (circa 1870 by Van Kesteren)

Dutch imported slaves and convicts to work the plantations, creating vast riches for the Dutch East India Company. Whatever your opinion of the modern multinational corporation may be, it is an undeniable fact that the institution's early roots were nourished by the blood of human beings, first in the Spice Islands and then in the slave plantations of the Caribbean, the American South, and other tropical locales that happened to be cursed with good soil, regular rains, and abundant sunshine.

The brutality of the Dutch regime—and their determination to maintain a complete monopoly on the world's supply of cloves and nutmeg—would eventually come back to haunt them. By the mid-

1620s, the Dutch had negotiated and slaughtered their way to complete control of the islands where these lucrative plants grew. But that dominance eventually provoked more than a few rivals to investigate the possibility of growing those crops in other locales, free of Dutch hegemony. And that seedling of an idea would eventually give rise to one of the greatest acts of commercial espionage in the history of capitalism.

There is some debate among historians over whether the French missionary, botanist, and master smuggler Pierre Poivre (literally "Peter Pepper") was in fact the original Peter Piper who, legend has it, picked a peck of pickled peppers. Certainly his name suggests a direct link (*piper* is Latin for "pepper"), and while pickled peppers do not appear in his biography, he did manage to pickpocket other valuable spices from the epicenter of the Dutch monopoly. Born in 1719, Poivre led a globe-trotting, multidisciplinary life. He served as a missionary in the Far East; sailed with the French East India Company; wrote a travel memoir and botanical survey that influenced Thomas Jefferson; and served as administrator of the French islands Isle de France and Bourbon, now known as Mauritius and Réunion.

At the age of twenty-six, Poivre was struck by a musket shot in the middle of a naval skirmish with the British; gangrene necessitated that most of his right arm be amputated, and he spent months recovering in the Dutch colony of Batavia, now Jakarta. His convalescence gave him plenty of time to observe the Dutch spice monopoly up close. His horticultural expertise—and his nationalist desire to help the French economy—put the idea in his head that with the proper care, clove and nutmeg plants could be cultivated in other parts of the world with climates similar to that of the East Indies. On his return trip to France, he spent time on the Isle de France and on Bourbon, and found both islands resembled the tropical rain forests of the Bandas and the Malukus. He would later write, "I then realized that the possession of spice which is the basis of Dutch power in the Indies was grounded on the ignorance and cowardice of the other trading nations of Europe. One had only to know this and be daring enough to share with them this never-failing source of wealth which they possess in one corner of the globe." He began sketching out a plan to "liberate" the seeds of clove and nutmeg from Dutch control, a plan that would make him almost certainly the most successful one-armed bandit in history.

Poivre's plan was both daring and extremely dangerous. The Dutch had grown out of some of their more bloodthirsty practices by this point, but their spice monopoly had been worth billions to their economy in modern dollars. They were not likely to hand over the seeds behind that empire to a Frenchman without a struggle. Poivre set sail across the Indian Ocean in 1750, arriving in the spring of 1751 in Manila, where he almost immediately encountered smugglers who sold him a supply of nutmeg seeds fresh enough that he could plant them. Thirty-two plants successfully sprouted, an early triumph that would keep him motivated through the many fruitless years that were to follow. But the nutmeg plants were only half the prize. Because clove spices are flower buds, not seeds, Poivre needed some way to get to an actual clove tree if he was going to introduce both spices to French soil.

At this point, Poivre began an almost comical split-screen existence: frantically trying to secure passage to the Malukus to steal the billion-dollar asset from under the noses of the Dutch, and at the same time lovingly maintaining his personal spice garden of nutmeg plants, which slowly began dying off. (One theory holds that the seeds had been poisoned by an operative of the Dutch East India Company, the horticultural version of the KGB putting plutonium in a rival spy's drinking water.) Poivre first tried to persuade the Spanish to take him to the Spice Islands, but they turned him down, not wanting to alienate the Dutch. Eventually he hired two

Pierre Poivre

boats and a Malay captain to take him to the Malukus, but naval battles in the southern islands delayed them so long that they ran up against monsoon season and had to turn back. By the time Poivre finally gave up on the cloves, in February of 1753, his nutmeg supply had dwindled down to nineteen plants. Stopping over on Pondicherry before rounding the horn of India, he reported only twelve still living. Somehow, five of the nutmeg sprouts survived the last thousand-mile journey back to safe harbor on Mauritius.

Once there, Poivre had to decide the best strategy for safeguarding his botanical treasure. "I knew from experience the incapacity of the gardeners of the Company at Mauritius," he later wrote, "and the little care they had given to the plants of all varieties which I had brought from the Cape and from Cochinchina. Most of the plants had either been dug up by neighbors or had died from neglect." Instead he quietly bequeathed the nutmeg plants to several friends on the island, planting them in three private gardens for extra protection. Before long, he was back in the East Indies, this time on a ship called the *Colombe* that managed to make it all the way to the Spice Islands. But this trip, too, proved to be ill-fated. He had difficulty anchoring off several of the islands; at one point, the ship's surgeon snuck off in a dinghy to warn the Dutch, though he was apprehended before he could do any harm. Poivre's attempts to woo the plantation workers and other non-Dutch residents were rebuffed. After months, he abandoned the scheme, and sailed west empty-handed.

Poivre's mission might have died there and left him an obscure footnote in the history of spice, had he not decided to write his memoirs, which he eventually published as *Voyages d'un Philosophe*. Thomas Jefferson read the book and hatched a scheme to bring rice farming to the American south based on Poivre's description of Chinese rice farms. A French minister back in Paris also read the memoirs with great interest. Inspired by Poivre's vision of bringing valuable spices—and profits—to the island colonies of the French empire, the minister arranged a position for Poivre as administrator of Isle de France and Bourbon, and secured official support for the plot to grow Dutch spices in French soil. In 1770, Poivre dispatched two ships to the Spice Islands, led by captains with long experience in the region. They returned with thousands of healthy nutmeg sprouts and three hundred clove seedlings. The first cloves to grow anywhere outside a hundred-mile radius of Indonesia were harvested in 1776 on Isle de France. It seems preposterous to say it, but one of the key events that brought an end to the Dutch financial empire of the seventeenth and eighteenth centuries was a one-armed Frenchman stealing a handful of seeds halfway around the globe.

Poivre's triumph, in a way, marked the beginning of the end for the spice trade, or at least the spice trade in its grandiose phase. Before long, descendants of Poivre's initial batch of clove seedlings would be employed to launch a thriving clove business in Madagascar and Zanzibar. "Nearly two hundred years later," Turner observes, "the flow of spices across the Indian Ocean has been reversed, with Indonesia now a net importer of cloves." The newfound geographic diversity of cloves, nutmeg, cinnamon, and pepper—and the increasingly efficient means of harvesting them—soon demoted spices from luxury item to commodity.

Interestingly, one of the last spices to descend the ladder into commodity status originated not in the Far East but in the Americas. On the gulf coast of Mexico, in a region now known as Veracruz, a vine with distinctive pale yellow flowers grows in the tropical forests. The vine is, technically, a member of the orchid family, but it is unique among the more than twenty-five thousand varieties of orchids in that it produces a crop: the vanilla bean. After pollination, the ovaries of the plant expand into long, thin fruits containing a multitude of black seeds. Thousands of years ago, the Totonac people of the Veracruz region discovered a technique for drying and curing these fruits that released a tantalizing aroma, a scent that would eventually suffuse ice cream parlors and birthday parties around the world.

"Perhaps the earliest known use for vanilla pods was as a simple but effective deodorant for the Indian's houses," historian Tim Ecott writes, "and it is still used in that way in central Mexico today, where a bunch of dried beans is tied together and suspended with string from a hook on a wall. Traditionally, the Totonac women, and women from other tribes in whose territory the plants grew, would place oiled vanilla beans in their hair, perfuming it with the subtle scent from the plant." By the time Europeans arrived in the Americas, vanilla had become an important, if luxurious, part of the Aztec culture. The elite ground up the pods and used them to take the edge off the bitter taste of the chocolate drinks they made from the cacao plant. Regions that produced vanilla often paid their taxes to the Aztec state in vanilla pods.

Today, vanilla has come to be imagined as the yin to chocolate's yang, an opposition heightened by the entirely arbitrary light coloring of vanilla ice cream and cake, but the truth is vanilla crept its way into the European palate as a kind of chocolate enhancer. A 1685 treatise entitled *The Manner of Making Coffee, Tea, and Chocolate* noted of hot chocolate drinks, "Everyone uses this confection, and puts therein Three little Straws or as the Spaniards call them Vanillas de Campeche. Our Vanillas are used in making the Chocolate, the which are very pleasant to the sight, they have the smell of Fennel, and perhaps not much different in quality, for all hold that they do not heat too much, and do not hinder the adding of Annis seed." Eventually vanilla broke free of its codependence on chocolate. Thomas Jefferson developed a taste for the spice during his years in Paris, and returned to the states with a handwritten recipe for vanilla ice cream, one of his many enduring contributions to American culture. In 1791, Jefferson sent a note to the American emissary in Paris asking for special assistance in locating vanilla beans to be shipped back stateside, along with cases of fine Bordeaux wines: "[My butler] informs me that he has been all over town in quest of Vanilla, & it is unknown here. I must pray you to send me a packet of 50 pods (batons) which may come very well in the middle of a packet of newspapers. It costs about 24 sous a baton when sold by the single baton." Vanilla never incited the mass obsession that erupted around pepper

Vanilla orchids and colocynth

or psychoactive compounds like coffee, tea, and chocolate. But it was flavorful enough—and, crucially, scarce enough—that, by the middle of the 1700s, pods of vanilla were worth their weight in silver.

That scarcity resulted from a strange twist of evolution. The orchid species that produces the vast majority of vanilla consumed in the world—*Vanilla planifolia*—can only be fertilized, in nature, by one species of bee native to Mexico and parts of Central America; the reproductive organs of the plant are so carefully guarded that other bees and insects that haven't coevolved with the flower will almost never accidentally fertilize it just by bumbling around in search of nectar and swiping some pollen in the process. In a sense, *Vanilla planifolia* evolved a kind of combination lock in the design of its petals that only one specific insect can get past. For centuries, the complexity of that system confounded humans as well. After Cortés and his men brought back word of vanilla to Europe, cuttings of *Vanilla planifolia* were successfully planted in tropical locations (and even Northern European hothouses) all over the world. Like most orchids, the vine was lovely to look at, but without the Mexican bees, the plant stubbornly refused to bear fruit. The Dutch may have guarded their clove and nutmeg monopoly with warships and genocide, but the Mexican monopoly on vanilla was guarded by the petals of a flower.

The story of how the lock that protected the treasure of *Vanilla planifolia* was eventually picked may be the ultimate example of the way the spice trade bound the planet together in a network of unlikely affiliations. The story begins where we left off with Pierre Poivre—on the French islands of Isle de France and Bourbon. Poivre's dream of replicating the lucrative crops of the Bandas on Isle de France and Bourbon hadn't entirely come to pass in the years that followed his daring thieveries in the East Indies. (The clove plantings, for instance, never really took off until they made it to Madagascar.) But Bourbon proved to be a useful harbor for the French sailing back from India and the East Indies, in part because the island had been entirely uninhabited when Europeans first arrived on its shores. Over time, more than a thousand species of plants from around the world were cultivated on the islands; slaves were brought in from Africa to work coastal plantations where coffee, sugar, and cotton generated handsome profits for the French empire. The names of the islands changed as the political regimes shifted back in Paris: after the House of Bourbon collapsed, Bourbon was christened Isle de la Réunion; it spent a decade or so as Isle Bonaparte before reverting back to Bourbon during the Restoration. Only after the 1848 revolution did it finally settle on its modern name: Réunion.

Vanilla plants had been introduced to Réunion several times before that last name change; a bundle of cuttings that arrived from Paris in 1822 were dispersed through a number of plantations. The vines grew for as long as two decades, and would flower occasionally. But, deprived of the deft touch of the Mexican bees, the plants would almost never produce fruit. All that would change, though, thanks to the ingenious horticultural explorations of a twelve-year-old boy: Edmond Albius, a slave who worked on a plantation known as Bellevue. Albius had hit upon a method of fertilizing the plant, a delicate maneuver that involved opening up the lip of the flower with his thumb and using a stick to press two parts of the flowers' reproductive organs together. As his master and surrogate father Ferréol Bellier-Beaumont would later recall, "This clever boy had realized that the vanilla flower also had male and female elements, and worked out for himself how to join them together." Albius's technique—"*le geste d'Edmond*," as it came to be called—soon spread across the island. Before long, the plantations of Réunion were shipping cured vanilla pods by the ton. Within a half century of Albius's discovery, the small island produced more vanilla than all of Mexico.

While the French prospered mightily from *le geste d'Edmond*, Edmond himself did not fare as well. Liberated in 1848, he was arrested and imprisoned for jewel robbery several years later, though his sentence was commuted as an acknowledgment of his contributions to Réunion's economy, after intense lobbying from Bellier-Beaumont. He died in poverty on the island in 1880. The local paper recorded his demise with little sugarcoating: "The very man who at great profit to this colony, discovered how to pollinate vanilla flowers, has died in the public hospital at Sainte-Suzanne. It was a destitute and miserable end."

Edmond's is the story of spice in a nutshell, a story where plants and peoples from all across the globe are tossed together—sometimes in great triumph and sometimes in great tragedy—all to take the rarest of tastes and turn them into commodities. A plant indigenous to Mexico and controlled by the Spanish is planted on an island in the Indian Ocean by the French, where it is first fertilized by a boy whose African ancestors had been brought to the island by French slave traders. And that seemingly trivial act—a boy tricking a flower into producing seed, in the hills of a remote island—would somehow shift billions of dollars of economic activity from one part of the world to another, and turn a spice that was once pursued by only the elites of society into a flavor so ubiquitous that its name has become a synonym for the commonplace and the ordinary.

Every grade school history textbook will tell you that the spice trade played a pivotal role in world history. But it is worth pausing for a moment to contemplate how many key developments and customs—many of which persist to this day—have spices at their origin: international trade, imperialism, the seafaring discoveries of Columbus and da Gama, the fall of Rome, joint-stock corporations, the enduring beauty of Venice and Amsterdam, global Islam, even the multicultural flavor of Doritos. Having a taste for spice is not just one of the luxuries that the modern world affords us; having a taste for spice is, in part, why we have a modern world in the first place. The most perplexing thing about that legacy is not the fact that spices were once fabulously expensive and are now cheap. (The pattern of luxury goods becoming mass commodities is in some sense the macro-narrative of capitalism: from cinnamon to cotton to computers.) The real question is why human beings were willing to pay so much money for such frivolous tastes.

The closest equivalent in modern times to the global impact of spices is our current appetite for oil. Just like the quest for spices, the quest for oil has compelled humans to redraw political borders, launch devastating wars, make brilliant scientific and technological breakthroughs, and create some of the most profitable companies in history. But at least fossil fuels, for all their faults, are actually *necessary* given the energy requirements of our lifestyles. You could easily imagine wars or empires or corporations being launched in the name of basic nutrition; food is life, after all. But conquering the world in the name of flavor? Where is the sense in that?

The conventional explanation for the spice craze attributes it indirectly to the need for basic nutrition. In Roman or medieval times, the story goes, the available food required massive amounts of spice to make it edible. Pepper and its ilk were considered a supplement to salt: a way of preserving food through the depths of winter, and of disguising the flavor of meat that had begun to spoil. This theory has been widely discredited for relatively straightforward reasons. The fantastic expense of spices like pepper or nutmeg meant that they were almost exclusively used by the European elite until the price began to fall in the 1600s. In medieval England, the extended royal family single-handedly dominated the market for spices. (Edward I, for instance, spent as much on spices annually as an earl would earn in a typical year.) The European aristocracy had no shortage of fresh meat or fish to consume, and they had plenty of salt to preserve anything that needed a longer shelf life. Spice was a craving, not a necessity. "To limit their function to food preservation and explain their use solely in those terms," the German historian Wolfgang Schivelbusch writes, "would be like calling champagne a good thirst quencher."

Were spices just a financial bubble, like Dutch tulips and Pets.com stock? Almost certainly not. To begin with, if spices were merely a bubble, it was the longest in the history of markets; it took almost two thousand years to burst. More importantly, the market price of spices like pepper and cinnamon was rarely influenced by second-order speculation: people driving up the price by betting that the price will go up. Those derivative markets wouldn't flourish until after the heyday of spices, partly because of economic institutions, like publicly traded companies, that spices helped invent. Spices were expensive because wealthy people were willing to pay significant sums to consume them, not just bet on their future value.

Why, then, were the medieval aristocrats so obsessed with spices? Consider an occupation that was once common in wealthy households, now lost to history: the speciarius, or spicer. According to Turner, Philip V of France had only four officers in his chamber: a tailor, a barber, a taster, and a spicer; the English king Edward IV had an entire "Office of Greate Spycerye." Tailors and barbers are still recognizable functions, and a taster makes intuitive sense given the threat of poisoning. But why carve out an entire office for the spicers? The word itself conjures up an image of some officious valet, standing at attention all day with a pepper grinder in his hands. But the medieval spicer provided much more than flavoring. For starters, managing an inventory of spices was itself a complex and challenging job, given their value and the scale of the royal feasts that featured them so prominently. But the spicer also played a crucial role that makes no sense in the modern context: he was somewhere between a pharmacist and a lifestyle coach. He was there to advise the royal family on the daily rhythms of their health and well-being: everything from chronic illnesses to sleeping patterns to bowel movements. And spices were the primary means of altering or preserving those rhythms.

Engraving of African medicinal plants, 17th century

Over its long history as a health supplement, pepper was considered a remedy for everything from cancer to toothache to heart disease. The Spanish king Philip II sent his personal physician to Mexico in 1570 to analyze the pharmacological potential of local plants. His description of the medical properties of vanilla alone suggests just how powerful these new compounds were considered to be: "a decoction of vanilla beans steeped in water causes the urine to flow admirably; when mixed with mecaxuchitl, vanilla beans cause abortion; they warm and strengthen the stomach; diminish flatulence; cook the humours and attenuate them; give strength and vigour to the mind; heal female troubles; and are said to be good against cold poisons and the bites of venomous animals."

Sexual dysfunction, too, was reliably treated with spice-based remedies. In *De Coitu*, the sex manual composed by the eleventh-century Benedictine monk Constantine the African, almost twenty recipes for sexual enhancement are listed; all of them rely on spices for their aphrodisiac powers. To the modern ear, the descriptions sound more like the ingredients for a delicious salad dressing than some medieval Viagra. Constantine describes an "electuary that I made for impotent men of cold and moist complexion: 6 drams ginger, camomile, anise, caraway; 4 drams hellebore seed, onion seed, colewort seed, ameos; 2 drams long pepper, black pepper, nut oil, as much honey as needed." As late as 1762, a German physician published a study in which he claimed, "No fewer than 342 impotent men, by drinking vanilla decoctions, have changed into astonishing lovers of at least as many women." Perhaps inspired by these accounts, the Marquis de Sade served copious amounts of vanilla with his desserts, to inflame the passions of his dinner guests.

Spices were not just considered to have medicinal attributes, in the soft, New Age way that we talk about the "healing properties" of aromatherapy today. In an age without Tylenol or antibiotics or Claritin, spices *were* medicine, running the gamut from what we would now call dietary supplements all the way to treatments for cancer or dementia. Much of this pseudoscience played off the then-dominant paradigm of bodily humors. Inherited from the Greeks, this framework conceived of health as a matter of balancing the four humors and their defining attributes: blood (warm and moist); bile (warm and dry); phlegm (cold and moist); and black bile (cold and dry). Unlike some folk medicines, the humoral system appears to have exactly zero correlation with any actual bodily mechanisms now known to science, but that didn't stop it from dominating the practice of medicine for almost two thousand years. The unhealthy body was a body in which one or more of the humors had a disproportionate influence, and that bodily harmony (or lack thereof) was heavily shaped by what you ate: pork and fish, considered cold and moist, would make you more phlegmatic; vegetables were considered dry, and thus exacerbated a bilious condition. Spices, then, were employed to keep the humors in equilibrium. Since many of them were considered hot and dry, they were often employed to combat the excesses of moist or cold dishes, especially meat, pork, and fish.

Actual medicines, delivered separately from meals, could take on astonishing complexity. These "compounds," as they were sometimes called, were prescribed by spicers to combat everything from gas pains to insomnia to depression. The most famous of these was theriac. Allegedly first concocted by the Persian king Mithridates VI as an antidote to reptilian venom, theriac was the ultimate panacea, used to treat—or ward off—just about any ailment possible. Some recipes included as many as a hundred separate ingredients. Measured purely by the complexity of its component parts, modern Doritos have nothing on theriac. A Dutch apothecary's guide from the late 1600s included both black and long pepper, cinnamon, nutmeg, and clove, along with more than fifty other ingredients—lesser calamint, white or common horehound, West Indian lemongrass, wall germander, Cupressaceae, bay laurel, and so on.

On the one hand, you can see in the staggering multiplicity of the theriac recipe the antecedent of modern miracle drugs, whereby billions of dollars and thousands of individuals converge to create a complex molecular cocktail designed to preserve our health. On the other, it seems bizarre to have such a meticulous prescription with so little actual medical value. And this helps us understand one underlying cause behind the strange prominence of the medieval spice: pepper, clove, nutmeg, and cinnamon became intensely valuable because the channels of innovation do not always run at the same speed. We invented a whole host of institutions and conventions that would ultimately turn out to be extremely useful in improving our health: diets and cookbooks shaped by a complex understanding of bodily systems, chemical compounds designed to treat illness and prescribed using standardized systems of measurement, printing presses and pharmacists that could disseminate those prescriptions. These were all significant innovations, not easily established. But as it happened, they arrived *before* the invention of the scientific method, randomized double-blind control drug trials, and other regulatory mechanisms that separated the genuine healers from the charlatans. On some basic level, the medical properties of the spices were pure fantasy. But that fantasy, for all its absurdities, established a regimen of health and improvement that has carried on into modern life with better success. The spicers were a kind of trial run for the modern pharmacist and drug company, offering imaginary cures. We figured out the form for maintaining a healthy lifestyle before we invented a method for scientifically testing the content. In *Middlemarch*, George Eliot described this process, evoking the history of alchemy and its delusions, but it could just as easily be applied to the sham medicine of the spicers: "Doubtless a vigorous error vigorously pursued has kept the embryos of truth a-breathing: the quest of gold being at the same time a questioning of substances, the body of chemistry is prepared for its soul, and Lavoisier is born."

Mostly these folk remedies were harmless, medically speaking; given the placebo effect, they might well have had a slightly beneficial impact. But, at least once, the use of spices as medicine seems to have backfired in a truly catastrophic way. The aromas of Oriental spice were said to combat the miasmatic air that conveyed plague. An Oxford fellow named John of Eschenden recommended "a powder of cinnamon, aloes, myrrh, saffron, mace, and cloves" to ward off the Black Death. A century later, an Italian endorsed Eschenden's prescription, calling it a "marvellous medicine against the corruption to the air in the time of pestilence." The carrier for the bubonic plague was the black rat, with the memorable Latin name *Rattus rattus*, whose fleas ultimately transmitted the disease to humans. *Rattus rattus* is not native to Europe; the species originated in Southeast

Asia and almost certainly spread to Europe via the channels of the spice trade during the late Roman Empire. By medieval times, the rodent had proliferated across Europe, thriving in its dense and polluted city centers. In January of 1348, according to the Flemish writer De Smet:

> *Three galleys put in at Genoa, driven by a fierce wind from the East, horribly infected and laden with a variety of spices and other valuable goods. When the inhabitants of Genoa learnt this, and saw how suddenly and irremediably they infected other people, they were driven forth from that port by burning arrows and divers engines of war; for no man dared touch them; nor was any man able to trade with them, for if he did he would be sure to die forthwith. Thus, they were scattered from port to port.*

Within a matter of months, the continent was under siege from the plague bacillus. Europeans, as it turned out, had it exactly wrong about spices. They weren't protection against the Black Death. They were the reason the Black Death came to Europe in the first place.

The folk medicines peddled by the spicers should not lead us to a purely utilitarian explanation for the spice trade, however illusory its cures may have been. Another, more subtle appeal shaped the European craving for spices. To eat nutmeg or cinnamon or pepper two thousand years ago was, in a real sense, the most tangible way to experience the mystical realms that lay beyond the edges of the map, what Europeans would eventually call the Orient. Remember that, during the height of the Roman obsession with pepper, accurate maps of India didn't exist, and the cloves that arrived from the Spice Islands were traveling from a part of the world that no Roman had visited. The first true globe-trotters were not human beings; they were the seeds and berries and bark of plants that were passed along the global relay system, from the Far East to the banquets of Rome. No wonder there was something magical about experiencing these flavors: you couldn't *see* the Orient in photographs or television images; you couldn't even point to it on a map. But you could taste it.

The Eastern allure was particularly powerful because a rich literature had developed suggesting that these distant parts of the world were, literally, an earthly paradise, either the actual location of the Garden of Eden or some close approximation. These origins help explain why spices were considered to have such significant medical powers; you were consuming something that had originated outside the fallen, debauched state of civilized man. To savor the clove or the nutmeg was to retreat back to a prelapsarian state, to eat as Adam or Eve had before their fateful encounter with the apple tree and the serpent.

However beautiful the Spice Islands may be, most of us would now consider them to be paradise with a lowercase *p* and not an actual Garden of Eden. But the vast distance those spices traveled does raise an interesting question. Almost all of the spices that eventually dominated the world's cuisine originated far from Europe, mostly in Southeast Asia: pepper in India; clove, nutmeg, and cinnamon near Indonesia. Columbus was famously wrong about cinnamon and pepper in the Americas, but South America did ultimately contribute the chili pepper and vanilla to the global spice cabinet. If we are trying to explain how the world system of international trade came about—with both the cosmopolitan intermingling of cultures that we rightfully celebrate, and the horrors of slavery and imperialism that we rightfully denounce—then one of the central questions we have to ask is this: Why did Europeans have to travel so far to find spices? Imagine an alternate scenario in which pepper grows naturally in Spain, and cinnamon abounds in France, and clove trees dot the foothills of the Italian Alps. The course of human history would likely be completely redirected: Europe remains far more insular; Columbus and da Gama never bother to set off in search of a direct route to the East. Without the immense markup of a global spice network, the accumulated wealth of Venice and Amsterdam and London dissipates, along with all the pioneering works of art and architecture that wealth funded. Without a vibrant pepper trade with India, calico fabrics never make it to the drawing rooms and garment stores of London; without a booming market for cotton textiles, the industrial revolution is delayed for decades.

The extent to which the spice trade had bound the globe together was perceived sharply by many of the participants—even those who never boarded a vessel and set sail for the Far East. After commissioning the East India Company in 1600, Elizabeth I wrote a handwritten letter to the "Great and Mighty King of Aceh," who controlled much of the pepper markets that had prospered around Sumatra (now Indonesia) in the 1500s. The missive was delivered in person by a British naval hero named James Lancaster, who led a small fleet of East India Company ships that arrived in Aceh in 1602. Elizabeth's language is remarkable in its supplication; she talks a great deal about the "love" between her nation and Sumatra, no doubt trying to distinguish the British from those rapacious Dutch and Portuguese. But perhaps the most extraordinary passage comes when she attempts to integrate the global reach of the spice trade into a broader story of Divine Purpose: God, she explained, saw fit to distribute the "good things of his creation . . . into the most remote places of the universal world . . . he having so ordained that the one land may have need of the other; and thereby, not only breed intercourse and exchange of their merchandise and fruits, which do so superabound in some countries and want in others, but also engender love and friendship between all men, a thing naturally divine."

Knowing the grim history that was to follow—almost four hundred years of colonial exploitation and slavery in pursuit of those "good things"—it is hard not to be cynical, if not outright appalled, by the talk of "love and friendship between all men." But Elizabeth

Clove harvesting

does hit upon an essential fact: that spices were distributed into the "most remote places of the universal world." The taste for those spices compelled human beings to invent new forms of cartography and navigation, new ships, new corporate structures, not to mention new forms of exploitation—all in the service of shrinking the globe so that pepper raised in Sumatra might more efficiently be delivered to the kitchens of London or Amsterdam. The automata of Merlin's Mechanical Museum demonstrated how the pursuit of leisure and play can be, on a conceptual level, exploratory, driving the creation of new social customs, materials, technologies, markets. But the pursuit of spice was *literally* exploratory: those strange new flavors propelled human beings around the globe like nothing that had ever come before them. Today's global village has its roots in the frivolity of spice.

This still doesn't answer the question of *why* Europeans had to travel so far, assuming we aren't satisfied with Elizabeth's providential explanation. Why weren't the Spanish hills teeming with pepper vines? Here the ecosystems approach to human history, most famously presented in Jared Diamond's *Guns, Germs, and Steel*, offers the most enlightening explanation. Diamond argued that civilization first took root in Mediterranean climates because those climates featured short rainy seasons and long dry seasons, which encouraged the cultivation of large-seeded grains like wheat and barley that became central to modern agriculture. Large, dense human settlements have become eminently possible outside of Mediterranean climates; indeed, some of the largest cities in the world are now located in Southeast Asia and South America. But it is very difficult to *invent* a large, dense settlement for the first time in, say, a tropical rain forest. And so the first societies that clustered into proto-cities, and left behind the subsistence lifestyles of the hunter-gatherers, were almost all located outside the tropics, in places rich with large-grained cereals.

Yet the challenges tropical climates pose for inventing civilization turn out to be opportunities when it comes to *biological* invention. There are more interesting flavors in the tropics because there is, on some basic level, more of everything in the tropics; that's why anyone who values biodiversity is so distraught by the destruction of the rain forests. Tropical plants evolve such a wide array of chemical compounds in their fruits and berries and seedlings because they are interacting with a much wider array of parasites and predators and coevolutionary partners than their equivalents in the Mediterranean climates of Spain or Italy. Most of those chemical innovations turn out to be useless to humans, but some small portion turn out to have value, as medicines, materials, intoxicants, or just interesting flavors.

In his complaint about the Roman dependence on imported spices, the historian Pliny observed, "It is remarkable that [pepper's] use has come into such favor: for with some foods it is their sweetness that is appealing, others have an inviting appearance, but neither the berry nor the fruit of pepper has anything to recommend it. The sole pleasing quality is its pungency—and for the sake of this we go to India!" We now understand the biochemistry of that pungent taste. Black pepper contains a molecule called piperine that activates a set of receptors on the surface of our tongues known as the transient receptor potential channels. These "TRP" receptors evolved to detect potentially harmful substances making contact with the skin; they function as a kind of alarm system for the body. When you accidentally grab a searingly hot pan, TRP receptors convert a chemical reaction happening at the epidermal layer into an electrical nerve signal that creates, in the mind, a sharp sensation of pain in your fingers. Both piperine and the active ingredient in the chili pepper, capsaicin, activate the TRP channels. This is why, on a biochemical level, it is not just a metaphor to say that spicy foods are "hot," even if they are served at room temperature. The piperine or capsaicin creates a kind of illusion of heat, triggering the same bodily alarm that rings when you step on a hot coal by mistake.

Piperine and capsaicin almost certainly evolved in response to the biodiversity of their tropical homeland; from the plant's point of view, in a world teeming with organisms that might potentially consume your seeds, it makes sense to arm those seeds with a substance that makes another organism feel as though its mouth were on fire. (Fruits evolved a different strategy, creating seeds that could survive digestion, and ensuring a wide distribution by wrapping those seeds in a delicious sugary flesh.) The pungency of pepper that Pliny decried originally served as a kind of chemical weapon, threatening to burn any creature that dared to eat the plant's berries. The story of pepper is thus a kind of inverted version of Diamond's account in *Guns, Germs, and Steel*: civilization took root in Mediterranean climates because large-seeded grains of wheat and barley were nearby and easy to consume; the global market of the spice trade arose because pepper grew only in distant tropical climates whose biodiversity had made it advantageous for the pepper berries to be painful to consume.

The bioweapon strategy did not just evolve among the plants. In 2006, a team of researchers announced that they had, for the first time, demonstrated that venom from a tarantula native to the West Indies activated the same TRP receptor that piperine and capsaicin trigger. At some point in the distant past, the ancestors of that tarantula hit upon a survival strategy that involved simulating intense, painful heat with their venom. Today, the Frito-Lay corporation sells billions of dollars' worth of snack foods that rely on the exact same biochemical channels to create the perception of heat. When you taste that Doritos chip, you are receiving a signal that evolution has been crafting for millions of years, a signal with a simple, universal message: "Fire!"

And here's the amazing thing: we took that signal and turned it into something enjoyable and unthreatening, something we eat for fun. Our genes want us to be wary of compounds that activate the TRP receptors, but we are not slaves to our genes. Sometimes the patterns and conventions of human culture flow naturally out of our evolutionary heritage, as with marriage rituals, spoken languages, and incest taboos. But often the true yardstick of cultural innovation comes in measuring how far the habits and customs and appetites of culture have taken us from our genes. Like many forms of delight, the taste for spice propelled us far from our roots—not just geographically but also existentially. That strange new taste on the tongue that would send any child into howls of pain could be savored by an adult, its pain turned into pleasure. Spices enlarged the map of possible desires, which in turn enlarged the map of the world itself. This boundary pushing, this constant reimagining of what our needs and appetites should be, may not be a "thing most divine," as Queen Elizabeth had it. But it is what makes us different from most organisms. What makes humans human is, in part, their ability to expand the boundaries of what it means to be human. The exploratory need for new experiences, new desires, and new tastes is, more often than not, the force behind that expansion. You might even call it the spice of life.

4

Illusion

The Ghost Makers

Leipzig, 1771. A handful of nervous but adventurous young men are gathered at the doorway of a second-floor room above a coffeehouse at the corner of Klostergasse and the Barfussgasschen. They are out on the town for the night, taking in a new spectacle that has been much discussed among the pamphleteers and chattering classes of Northern Europe. After a few minutes of waiting, a hooded figure emerges from the darkened room and beckons them inside. Death's-head masks line the walls; an altar draped in black cloth stands across from them. The smell of sulfur pervades the space, illuminated only by a few flickering candles. Their host stands in a chalk circle at the center of the room and begins to read aloud ancient incantations conjuring the spirit world. At once, a blast of noise erupts in the room, and the flames are extinguished. In the gloom, a spirit appears, hovering over the dark altar. The guests feel a literal shock as the specter confronts them, a current of electricity running through their bodies. The host stabs at the conjured ghost with a sword to demonstrate its ethereal presence; its mouth opens and begins to speak, in "a hoarse and terrible tone."

This eerie spectacle was the creation of a young German named Johann Georg Schröpfer, a troubled showman who would for a brief period of time become one of the most famous men in all of Europe. In the mid-1760s, Schröpfer had taken a job as a waiter at a Leipzig lodge that was frequently used by local Freemasons for their ritualistic gatherings. He soon assimilated the Freemason dogma and began styling himself as a channel to the spirit world. Like many in the orbit of the Freemasons, Schröpfer lived in the murky middle ground between science and occultism; he was versed in the new technologies of magic-lantern projections and dabbled in chemistry. But he also had a fascination with séances, and seems to have believed that using technology to create the illusion of spectral appearances could actually enable contact with the spirit world.

The combination of showmanship and spiritualism would prove to be lethal. In 1769, Schröpfer purchased the Leipzig coffeehouse, and promptly retrofitted its billiards room into a multimedia, immersive theater of terror. Before long, he was conducting séances for his acolyte customers, using magic lanterns to project ghostly images against curtains of smoke, while creepy sound effects thundered in the small space. Anticipating the "Sensurround" gimmicks of 1950s Hollywood, like the Vincent Price camp classic *The Tingler*, Schröpfer delivered the electric shocks to his clientele using the static-electricity machines that had become popular parlor amusements during that period.

Schröpfer had stumbled across a form of entertainment that would eventually mature into the immensely profitable genre of horror films. Before long he had become a legend across Europe: the *Gespenstermacher*—or "ghost maker"—of Leipzig. Pamphleteers debated whether his conjuring was reality or illusion, but Schröpfer appears to have fallen under his own spell. In addition to his necromantic illusions, Schröpfer crafted an entirely fictitious persona for himself as a man of great means whose fortune had for some obscure reason taken the form of a hidden treasure guarded by bankers in

Frankfurt. The mix of occultism and constant deception eventually pushed an already disturbed Schröpfer over the edge. In 1774, he went for a stroll in a Leipzig park with a handful of friends, promising them "something you have never seen before." At one point in their sojourn, he walked around a corner out of the view of his companions, who were then startled by a loud explosion. When they caught up with their friend, they discovered him bleeding to death on the ground, the victim of a self-inflicted gunshot wound to the head. Clearly deranged, Schröpfer had vowed to return to life during a future séance. No horror auteur since has shown such dedication to the suspension of disbelief.

The ghost maker's shocking demise only heightened his reputation. "In dying," Deac Rossell writes, "Schröpfer became the Lautréamont, the James Dean, the Jimi Hendrix of his generation." While the debate continued to rage over the legitimacy of Schröpfer's black arts, dozens of showmen across the continent constructed stage shows that re-created his special effects. The most successful of those disciples was another mysterious German named Paul Philidor, who began staging horror shows in Vienna in the late 1780s. Unlike Schröpfer, Philidor made no pretense that he was actually conjuring up the dead. (He referred to himself as "The Physicist," presumably to differentiate himself from the ghost making of Schröpfer.) A newspaper report from 1790 described the show: "As soon as the peroration begins, distinct thunder is heard approaching, accompanied by wind, hail and rain; the lights extinguish themselves one by one, and in the impenetrable darkness various ghosts of all shapes flutter about the room; finally after a very furious storm and a rushing of the wind, a living ghost appears out of the Earth, then again slowly sinking into the abyss below."

Philidor made three essential contributions to the genre. First, he began rear-projecting the spectral images on a thin, semitransparent curtain that was otherwise invisible to the spectators. (Schröpfer's technique of projecting onto clouds of smoke was undeniably eerie, but also much less reliable as a canvas for the images—and filling a small chamber with thick smoke made the séances physically unpleasant.) Philidor also pioneered the technique of placing the magic lantern on wheels; by slowly moving the projector toward the screen, he created the illusion that the specters were growing larger as they approached the terrified spectators. (A little more than a century later, the technique would be reimagined as the cinematic tracking shot.) Philidor's final contribution was a linguistic one. He gave his spook show a name, one that would haunt the imagination of Europeans for decades to come: the Phantasmagoria.

The roots of the Phantasmagoria lay in German soil, but the show truly flowered in Paris. Philidor took his exhibition to the Hôtel des Chartres in 1792, at the height of revolutionary turmoil; throngs of amazed Parisians shivered at Philidor's illusions while Louis XVI stood trial before the National Convention. In April of 1793, while still performing for packed houses, Philidor mysteriously shut down his Phantasmagoria and went into hiding. (Rumors suggested he had somehow run afoul of the Committee for Public Safety.) A few years later, the Belgian showman Étienne-Gaspard Robertson revived the Phantasmagoria to great success, conducting his spectral displays in the vaults of an abandoned Capuchin monastery beneath the streets of Paris. The shows were a kind of fusion of two modern senses of terror: the audiences screamed and recoiled just as audiences at modern horror films do, but the ghosts Robertson conjured belonged to the political Terror, as magic-lantern projections of the death masks of Robespierre, Marat, and Louis XVI loomed in the shadow light.

New technologies or forms of popular entertainment change the world in direct ways: creating new industries, enabling new forms of leisure or escapism, sometimes creating new forms of oppression or physical harm to the environment. But they also change the world in more conceptual ways. Every significant emergent technology inevitably enters the world of language as a new metaphor, a way of fram-

The Phantasmagoria

ing or illuminating some aspect of reality that was harder to grasp before the metaphor began to circulate. As a show, the Phantasmagoria might seem to have been a folly, the eighteenth-century version of a slasher flick. But as a metaphor, it turned out to have a powerful philosophical allure.

Hegel invoked Philidor's creation in his Jena lectures, delivered while writing *Phenomenology of Spirit*: "This is the night, the inner of nature that exists here—pure self. In phantasmagorical presentations it is night on all sides; here a bloody head suddenly surges forward, there another white form abruptly appears, before vanishing again. One catches sight of this night when looking into the eye of man—into a night that turns dreadful; it is the night of the world that presents itself here." Schopenhauer described the human sensory

Double lens magic lantern

apparatus as a "cerebral phantasmagoria." In his 1833 satiric novel, *Sartor Resartus*, which contained a thinly veiled caricature of Hegel himself, Thomas Carlyle popularized the use of the term as a metaphor for an individual or society that has lost its grasp of reality: "We sit in a boundless Phantasmagoria and Dream-grotto; boundless, for the faintest star, the remotest century, likes not even nearer the verge thereof; sounds and many-colored visions flit round our sense . . ."

That broader definition of *phantasmagoria* as a kind of mass illusion would play a crucial role in one of the most influential paragraphs of political philosophy ever written, in the section of *Das Kapital* where Karl Marx first defines his notion of commodity fetishism:

> *As against this, the commodity-form, and the value-relation of the products of labour within which it appears, have absolutely no connection with the physical nature of the commodity and the material relations arising out of this. It is nothing but the definite social relation between men themselves which assumes here, for them, the fantastic form of a relation between things. In order, therefore, to find an analogy we must take flight into the misty realm of religion. There the products of the human brain appear as autonomous figures endowed with a life of their own, which enter into relations both with each other and with the human race. So it is in the world of commodities with the products of men's hands. I call this the fetishism which attaches itself to the products of labour as soon as they are produced as commodities, and is therefore inseparable from the production of commodities.*

"The fantastic form of a relation between things"—the English phrase derives from an 1887 translation, but the original German makes it clear that Marx was deliberately invoking the tradition of Schröpfer and Philidor: "the fantastic form" is in the original *"die phantasmagoriche Form,"* a "misty realm" with "autonomous figures endowed with a life of their own." Capitalism, to Marx, was not just a form of economic oppression but, crucially, an economic system in which the objects produced were wrapped in a kind of ghostly illusion, an unreality that kept its participants from recognizing the truth of their oppression. Marx was never reluctant to lean on arcane philosophical arguments or advanced economic theory to explain his vision of history: Hegel, Adam Smith, and David Ricardo are referenced dozens of times over the course of *Das Kapital*'s three volumes. But at a crucial point in that argument, introducing a concept that would prove to be one of his most influential, he drew upon popular entertainment, not phenomenology, to convey his meaning. New ideas need new metaphors, and in Marx's case the new metaphor came from a spook show.

In 1801, a man calling himself Paul de Philipsthal launched a new version of the Phantasmagoria at the Lyceum Theatre in the Strand. Many historians believe that Philipsthal was in fact Philidor himself, returning to the entertainment business with a new stage name, but Philipsthal's true identity has remained something of a mystery. The Phantasmagoria became an anchor tenant of the booming marketplace of illusion that flourished in the West End during the first decades of the century. At some point in this general time frame—the exact dates have been lost to history—the Scottish scientist David Brewster began frequenting the Phantasmagoria during his stays in London. Brewster is one of those nineteenth-century characters who have no real equivalent today. An ordained minister in the Church of Scotland, he took an early interest in astronomy and became for a time one of the world's leading experts on the science of optics. But he also harbored a great fondness for popular amusements. That obsession led him to invent the kaleidoscope, which was for a few years the PlayStation of the late Georgian era. (Either through incompetence or indifference, Brewster barely made a penny from the device, as imitators quickly flooded the market with clones of his original idea.) But his obsession also led Brewster to the spectacles of illusion and terror in the West End, to the Phantasmagoria and its brethren. He was there in part as a debunker, a skeptic discerning the secret craft behind the spectacle. But he also sensed that something profound was lurking in the trickery, that the showmen were exploiting the quirks in the human sensory system. Exploiting the quirks made them more visible to the scientist.

Of the Phantasmagoria's spectral images, Brewster compiled this analysis, accompanied by diagrams:

> *[The] phenomena were produced by varying the distance of the magic lantern AB, Fig. 5, from the screen PQ, which remained fixed, and at the same time keeping the image upon the screen distinct by increasing the distance of the lens D from the slides in EF. When the lantern approached to PQ, the circle of light PQ, or the section of the cone of rays PDQ, gradually diminished and resembled a small bright cloud, when D was close to the screen. At this time a new figure was put in, so that when the lantern receded from the screen, the old figure seemed have been transformed into the new one.*

Brewster's analysis didn't provide much ripe material for a promotional pull quote for the show. But then again, he was not reviewing the spectacles in the mode of a journalist or a critic. He was there at the encouragement of Walter Scott, taking notes for a book that he would come to call *Letters on Natural Magic.* Brewster had realized that, just as Enlightenment science had unlocked many doors for creating magical distortions of reality, it had also unlocked doors for detecting the laws behind that reality. The ability to understand the world advanced at roughly the same pace as the ability to deceive. That new deceptive power—*natural* magic, as opposed to the

David Brewster

supernatural kind—was at its most vivid in the West End of London. Brewster was there, at Scott's bidding, to explain away its illusions—and celebrate them at the same time.

The fact that a man like Brewster would seek out such popular entertainments is itself noteworthy. The "exhibitions" that began proliferating across Europe were among the first places where a shared culture began to take shape that compelled highbrow philosophers, common laborers, and landed gentry to take in the same amusements on more or less equal terms. Today we take it for granted that movie stars and politicians and factory workers will all happily turn out for the latest Pixar film or gather in a football stadium to cheer on their favorite team. But three hundred years ago, the different classes had few points of intersection. "At all times, curiosity was a great leveler," the historian Richard Altick writes in his magisterial work, *The Shows of London*. "Exhibitions that engaged the attention of the 'lower ranks' also attracted the cultivated and, perhaps to a lesser extent, vice versa. The spectrum of available shows was not divided into one category that was exclusively for the poor and unschooled and another, entirely separately for the well-to-do and educated. As more than a few foreign visitors noted, no English trait was more widespread throughout the entire social structure than the relish for exhibitions, and, one might add, no trait was more effective in lowering, however briefly, the conventional barriers that kept class and class at a distance." What brought them together was the strange, unpredictable pleasure of being fooled.

Early in *Letters on Natural Magic*, Brewster observes that a disproportionate amount of popular illusion is directed at the human visual system. "The eye," he wrote, "is the most fertile source of mental illusions . . . the principal seat of the supernatural." Since Brewster's time, entire books—some targeted at seven-year-olds, others at neuroscientists—have cataloged a vast menagerie of optical illusions. Consider the two famous visual tricks, the Kanizsa triangle and the Necker cube.

In each case, the eye detects something that is quite literally not there: a white triangle and a three-dimensional box. In each case, it is almost impossible to un-see the illusion. The Necker cube can be visually flipped between two different three-dimensional orientations, but most of us can't perceive it as it actually is: twelve intersect-

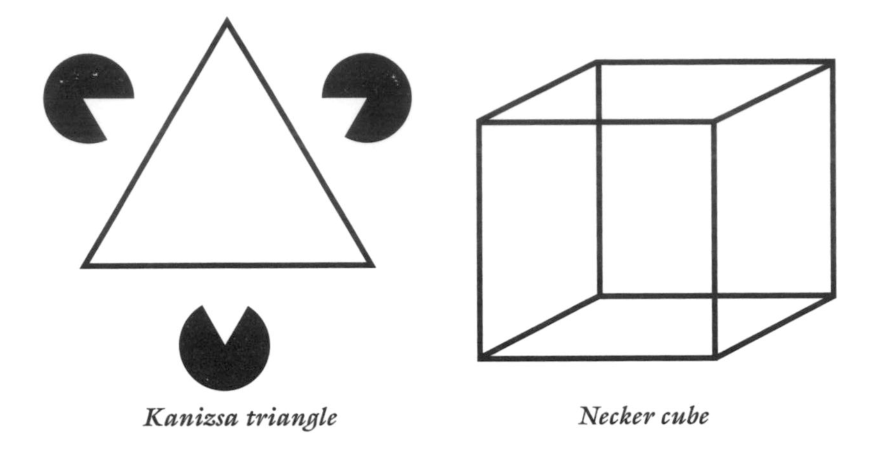

Kanizsa triangle *Necker cube*

ing lines lying on a two-dimensional surface. The mind's eye conjures up a perception of depth that empirically does not exist. The smallest tweak to the image can eliminate the 3-D effect, as in this drawing, known as a Kopfermann cube, where the image appears to alternate between a 3-D cube and a 2-D pinwheel shape.

Hundreds of similar illusions have been discovered by artists, showmen, and scientists over the centuries. Strangely, the human brain doesn't seem to be nearly as vulnerable to being fooled by the other senses. A few comparable auditory illusions exist, most famously the illusion created by stereophonic sound, which tricks the ears into perceiving a sound emanating from a point between the two speakers. A number of automaton designers in the eighteenth century attempted to create a "speaking machine"—a robotic human head that could utter words and sentences through artificial means, following on the principles of Vaucanson's auto-

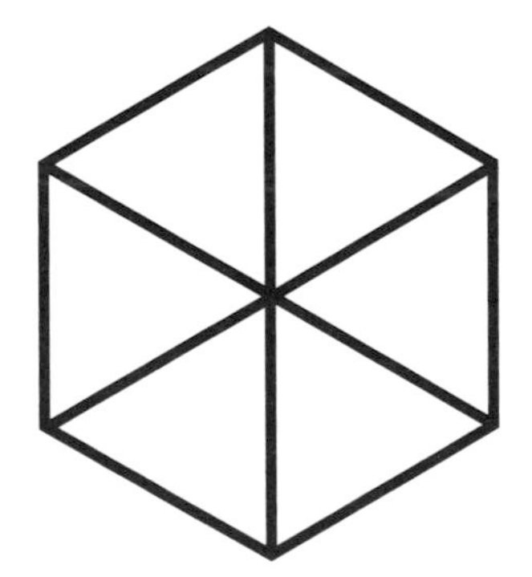

Kopfermann cube

mated flute player. But the human ear is not easily fooled by speech simulations: even today, with all of our computational power, a child can tell the difference between Siri and a human voice. And the other senses—touch, smell, taste—are even less prone to being tricked the way our eyes are tricked by the Necker cube. A handful of tactile illusions exist; with taste, the closest equivalent might well be the way chili peppers trick our brains into perceiving heat. But if you want to deceive the senses of another human being, your best bet is to do it through their eyes.

There is something paradoxical about this vulnerability. The human sense of sight is generally considered to be the most developed of our senses. Some estimates suggest that 85 percent of the information we take in arrives through our visual system. Why, then, should our strongest sensory tool be the most vulnerable to being tricked? Over the past few decades, researchers specializing in visual intelligence have come to understand this apparent paradox. The power of our visual system can't be measured in simple resolution, like the megapixels of digital cameras. In fact, the percentage of your visual field that you actually perceive in focused "high definition" is shockingly small. What makes our eyes so perceptive lies in the way the brain *interprets* the information it receives through the optic nerve. In a sense, the brain has evolved a series of cheats that enable it to detect things like edges or motion or three-dimensional relationships between objects, filling in missing information on the fly. You can think of these as the rules of thumb that govern our sense of sight. For instance, when our eyes perceive two lines coinciding in a flat image, our brain assumes those lines intersect in three-dimensional space. (The Necker cube relies on this rule to create the sense of depth in the image.)

Millions of years of evolution created rules for interpreting visual information, helping the eye evaluate and predict the physical arrangement and motion of objects that it perceives. But through hundreds of years of *cultural* evolution, we began discovering unusual configurations that would confound those predictions, forcing the eye to see something that wasn't, technically speaking, there. Natural selection created a kind of neural technology to interpret the information transmitted to our eyes, and then human beings deliberately set out to invent technology that short-circuited evolution's inventions. This turned out to be surprisingly fun.

Despite his scholarly expertise in the science of optics, Brewster himself was happy to defy the rules of perception for mass amusement. Late in his life he invented the stereoscope, the handheld technology that fools the eye into perceiving two distinct flat images as a single 3-D scene. Unlike his earlier kaleidoscope invention, Brewster managed to build a successful business selling his contraption, properly branded as a "Brewster Stereoscope." Queen Victoria famously marveled at one during the Great Exhibition of 1851. The stereoscope lives on to this day in the form of the popular View-Master toy, and the fundamental illusion the stereoscope relies on is also central to virtual reality goggles like Oculus Rift.

Optical illusions can be employed for more serious pursuits. Until the late nineteenth century, the most famous and influential "trick of the eye" was the invention of linear perspective, generally credited to the architect Filippo Brunelleschi, though the fundamental rules that governed the technique were first outlined in the book *On Painting* by Leon Battista Alberti. Like the Necker cube, it is almost impossible *not* to perceive the depth relationships in a painting that successfully executes the principles Brunelleschi and Alberti devised. Technically speaking, linear perspective is nothing more than an optical illusion, but it is rightfully considered one of the most transformative innovations of the Renaissance.

For a brief period at the end of the eighteenth century, it seemed as though a Scottish painter named Robert Barker had stumbled across an innovation of comparable significance. At some point in the mid-1780s, Barker took a stroll to the top of Calton Hill in Edinburgh. Standing near the current site of the Nelson Monument and gazing out over the city, Barker hit upon the idea of painting the entire 360-degree view by rotating a sequence of square frames around a fixed spot, sketching each part of the vista and then uniting them as a single wraparound image. With the assistance of his twelve-year-old son, Barker completed the epic project, but when he unfurled the final immense canvas and wrapped it around the viewer, he discovered that the concave surface distorted the image, making horizontal lines appear curved unless they were perfectly aligned with the viewer's eyes. In a sense, it was the opposite of the problem Brunelleschi and Alberti had solved: instead of creating the illusion of 3-D on a flat surface, Barker had to eliminate the distortions that came from painting on a 3-D surface. He ultimately devised a technique whereby straight lines would be artificially curved to compensate for the distortion, not unlike the way vanishing points bring parallel lines closer together in linear perspective. Barker also imagined an entire built structure to house his illusion, with concealed overhead lighting and an entrance through stairs below the viewing platform. (A doorway stuck in the middle of the painting would break the spell.) He was granted a patent in 1787 for "an entire new Contrivance of Apparatus . . . for the Purpose of displaying Views of Nature at large."

After successful prototype exhibitions in Edinburgh, Barker relocated to London, where he formed a joint-stock company, backed by a handful of wealthy investors, and began scouting for a site in the West End where he could produce his immersive spectacle to full effect. He sent his son to the roof of the Albion mills near Blackfriars Bridge to sketch the skyline of London, the way the two of them had captured Edinburgh from Calton Hill. At the suggestion of a "classical friend," Barker hit upon a name for his creation, drawing on the Greek phrase for "all-encompassing view." He called it the Panorama.

By 1793, Barker had constructed a six-story building near Leicester Square, designed for the exclusive purpose of displaying two sep-

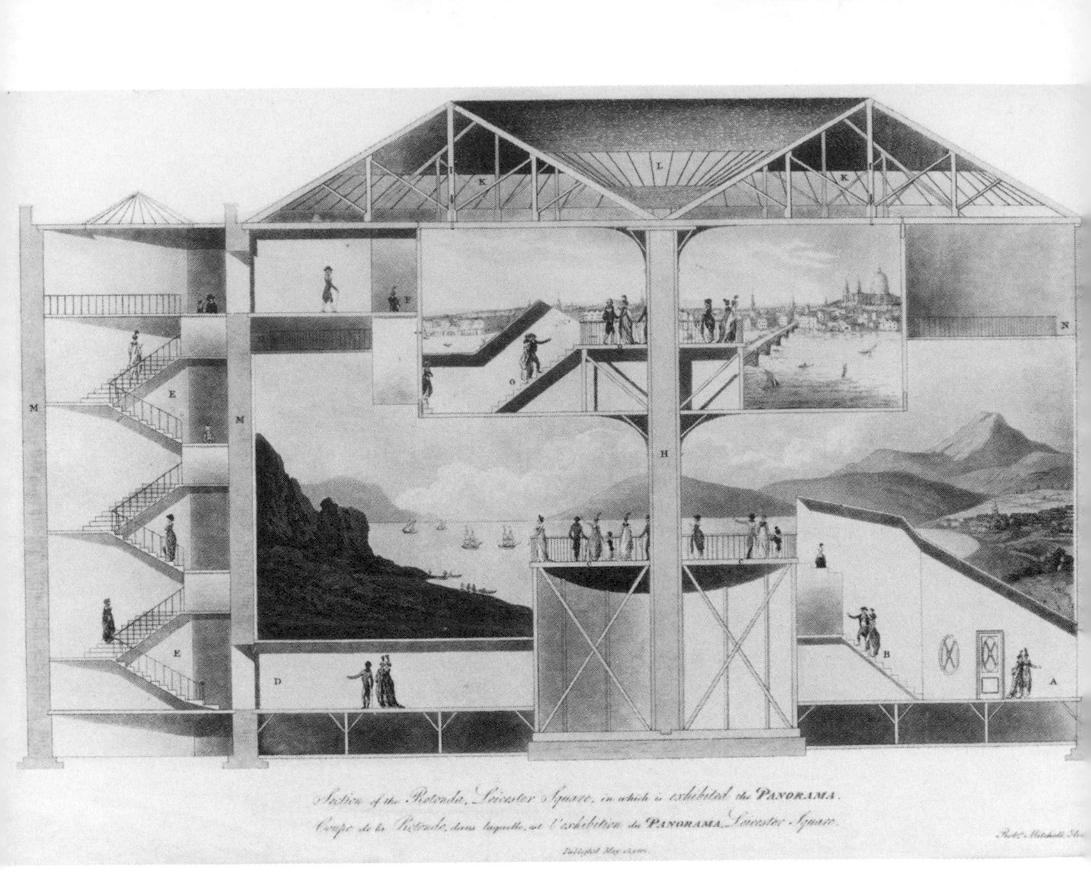

Cross section of the Rotunda in London where Robert Barker displayed two of his panoramas

arate Panoramas to crowds of paying spectators. The lead attraction was an immense vista of London, based on the Albion mill sketches, which encompassed 1,479 square feet. (A smaller Panorama simulated the British fleet sailing at Spithead, the bay near Portsmouth.) Barker ran advertisements that modestly suggested his technique was "the greatest IMPROVEMENT to the ART of PAINTING that has ever yet been discovered." For a time, the bombast seemed warranted. The show itself was a runaway success. The king and queen requested an advance viewing, though Queen Charlotte later reported that the illusion made her dizzy. The critical reception was equally enthusiastic: "No device . . . has approached so nearly to the power of placing the scene itself in the presence of the spectator," one observer wrote. "It is not magic; but magic cannot more effectually delude the eye, or induce a belief of the actual existence of the objects seen." Before long, imitations appeared across Europe, many re-creating famous military conflicts, some of them straight out of the news. The Panorama was the first of the temples of illusion to find a big audience in the newly formed United States. Coined specifically to refer to Barker's showcase, the word *panorama* entered the common lexicon of at least a dozen languages, signifying any kind of broad view, artistic or otherwise.

Appropriately enough, given the Panorama's strange brew of highbrow and lowbrow appeal, the best words describing its impact were penned by Charles Dickens decades after Barker opened the Leicester Square exhibit:

> *It is a delightful characteristic of these times, that new and cheap means are continuously being devised, for conveying the results of actual experience to those who are unable to obtain such experiences for themselves; and to bring them within the reach of the people—emphatically of the people; for it is they at large who are addressed in these endeavours, and not exclusive audiences . . . Some of the best results of actual travel are suggested by such means to those whose lot it is to stay at home. New worlds open out to them, beyond their little worlds, and widen their range of reflection, information, sympathy, and interest. The more man knows of man, the better for the common brotherhood among us all.*

Barker's Panorama suggested a higher purpose for the illusionists. Beyond the chills of the spook show or the amusements of the digesting duck, the illusion shows could also serve an almost journalistic function, reporting on current events. Journalists had been sending back dispatches from the front lines for at least a century before Barker staged his exhibits, but in an age without telegraphy and photography, those reports were slow in arriving and limited for the most part to textual accounts. (King George oversaw Great Britain's war with the American colonies with a four-week lag—the length of time it took news of the battles to travel by ship across the Atlantic.) The illusionists couldn't help speed up the transmission times, but they could marshal their powers to re-create the experience of being in battle, as Barker had done with the Panorama.

In September of 1812, the Bavarian musician and inventor John Nepomuk Maelzel found himself in Russia, just in time to witness the legendary burning of Moscow that greeted Napoleon's arrival in that city, and that would soon lead to his epic defeat there. The fire and the subsequent battle for Moscow would inspire many great works of art in the years that followed: Tolstoy's *War and Peace*, Tchaikovsky's *1812 Overture*. But one of the very first—and arguably most original—attempts to represent this world-historic event was engineered by Maelzel within a year of Napoleon's defeat, in the form of an animated diorama called *The Conflagration of Moscow*. Maelzel's creation premiered in Vienna, but he would ultimately take it across Europe and North America, dazzling audiences for decades with his mesmerizing reconstruction of the great city on fire.

A detailed inventory of the show that toured the United States gives some sense of the scale of the production. Movable frames representing the buildings of Moscow—the Kremlin, church spires, castles—were designed to collapse or explode on cue. Behind the skyline, Maelzel hung a transparent painting that suggested a haze of smoke and fire; behind it another painting depicted other buildings in the distance ablaze, with a moon glowing in the night sky above the carnage. At the front of the stage, two bridges and a causeway carried more than two hundred miniature Russian and French soldiers, featuring "musicians, snipers, cavalry, infantry mines and cannons"; the troops were pulled across the stage on hidden grooves, controlled by a hand crank. Fire screens enabled actual flames to creep across the urban landscape without actually damaging any of the equipment; Maelzel deployed small fireworks and burning pans of charcoal to enhance the effect. Lighting the tableau required "sixteen lanterns, twenty-five Argand lamps, six candlesticks with springs, snuffers, and trays, forty half-circular patent lamps with reflectors, nine square and six oblong lamps, and thirteen common japanned lamps with stands."

The experience of watching *The Conflagration of Moscow* was not, strictly speaking, *narrative* in its form. Yes, events followed a preordained sequence on the stage: Napoleon's army advancing; the Russians retreating; the flames surging across the skyline. But the true appeal of the spectacle came from the sense of immersion, just as it had with Barker's Panorama. "The city was before us, closely built up and the houses all aflame," one spectator recalled decades later. "We quivered at the sight; saw men, women and children making their escape from the burning buildings, with packs of clothing on their backs. The scene was terrible, and so realistic that when we went to bed after returning from the spectacle, we hugged each other and rejoiced that our house was not on fire." Through these elaborate illusions, the viewers lived through the sensory overload of one of recent history's most chaotic and terrifying events. In this, Barker and Maelzel had stumbled across an appetite in their audiences that would later be slaked by modern media's endless recycling of disaster footage, from the *Hindenburg* to 9/11. (Nonnarrative immersive simulations of famous catastrophes may well have a renaissance if virtual reality becomes a mainstream pastime.)

While the age of illusion was dominated by optical effects, thanks in large part to the evolution of the human visual system, Maelzel did more than any illusionist of his generation to explore the aural dimension of this mechanical theater, using a variety of bespoke instruments to simulate the sounds of a city under siege:

> *A musket machine had twelve springs to force striking hammers. Cannon drums were struck with the fist in a sparring glove . . .*

> *An "iron explosion machine" contained about a peck of stones, and on being cranked created a noise like the crash of falling buildings and the explosion of gunpowder. There were table bells, glass bells, and Chinese gongs to represent church bells and various other city sounds. Hand organs with bellows and a collection of cylinders provided the two types of martial music used . . . A trumpet machine, with twelve trumpets in it, was capable of playing a dozen tunes. A hand organ supplied the sounds of the cymbals and bass drums.*

Maelzel's attempt to re-create the dissonant music of war dovetailed with another of his projects, a jack-of-all-trades musical instrument called the panharmonicon that could simulate the sounds of an entire military band, all the way down to gunfire and the boom of the canons. The contraption, which was roughly the size of a large closet, controlled its instruments via an immense rotating barrel, a scaled-up version of the pinned cylinders that controlled the music boxes and automata of the preceding centuries. Keeping with the military theme of his illusions, Maelzel collaborated with a friend of his to compose a score explicitly for the panharmonicon, celebrating Wellington's victory over the French at the Battle of Vitoria. The friend happened to be none other than Ludwig van Beethoven. Alas, the panharmonicon never found a home in the classical mainstream. While Beethoven's piece—now known as "Wellingtons Sieg"—was performed by Maelzel's panharmonicon at a number of traveling exhibitions, the two men eventually had a falling-out, with Beethoven ultimately suing Maelzel and rewriting his composition for a traditional orchestra.

Maelzel's collaboration with Beethoven suggests just how difficult it is today for us to place the illusionists on the spectrum of high art and low amusements. From a modern vantage point, they seem like carnival showmen and hucksters. No doubt some of them were. Yet the cast of characters involved in this strange new culture—from Brewster to Beethoven—and the seriousness with which many of the most advanced spectacles were taken suggests that something more profound was under way. But wherever you place them on the spectrum of artistic expression, one thing is clear: by the first decades of the nineteenth century, the success of Barker's Panorama and Philipsthal's Phantasmagoria had set off a kind of entertainment version of the Cambrian explosion. Bizarre new species of illusion proliferated across the West End. (Smaller versions of this craze occurred in New York, Paris, and other cities as well.) The names themselves—with their strange Greek neologisms—suggest just how far the language was straining to represent the novelty of the experiences. Along with the Panorama and the Phantasmagoria, a visitor to London in the early 1800s could enjoy a "Novel Mechanical and Pictorial Exhibition" called the Akolouthorama; a predecessor to Philipsthal's spook show called the Phantascopia; an exhibition called the Spectrographia, which promised "TRADITIONARY GHOST WORK!"; an influential mechanical exhibition dubbed the Eidophusikon; the Panstereomachia, "a picto-mechanical representation," according to the *Times*. A virtual orchestra created by a painter and musician named J. J. Gurk entertained audiences with performances of "Rule, Britannia." (Confusingly, it was also called the Panharmonicon, though it had nothing to do with Maelzel's device.) Dozens of derivations of Barker's immersive paintings sprouted as well: the Diorama, the Cosmorama, the Poecilorama, the Physiorama, the Naturorama. An American showman named John Banvard popularized the "Moving Panorama," which simulating a ride down the Mississippi River by slowly unfurling a painting that was over a thousand yards long.

Some of these creations, like Banvard's scrolling landscape, were

Advertisement for an early slide projector, or "magic lantern"

genuine innovations; others were cheap knockoffs. (Of the Naturorama, the *Literary Gazette* sneered, "You are allowed to look through glasses at miserable models of places, persons, and landscapes; while two or three nasty people sit eating onions and oranges.") But what made the whole collection so remarkable was the sheer diversity of these wonderlands and magic shows. If you blur your eyes, the West End of London circa 1820 doesn't seem all that different from the West End of today, with its marquees promoting the latest hit comedy or Andrew Lloyd Webber revival. But the entertainment variety today lives within the formal conventions of theatrical plays: an audience gathers in an auditorium and watches actors perform scripted material on a lighted stage, sometimes accompanied by music. The variety of today's West End belongs to the content, not the form. Two centuries ago, there were plays and musicals, but there were also panoramas and magic-lantern ghost shows, and animated paintings populated by small robots—and dozens of other permutations. The West End functioned as a grand carnival of illusion, with each attraction dependent on its own unique technology to pull off its tricks.

Perhaps the single most significant fact about that carnival is this: almost every species in this genus of illusion died off by the dawn of the twentieth century. Many forms of entertainment from the end of the eighteenth century continue on in recognizable form. People still go to see musicals, attend operas, read novels, and visit art galleries. But other than the occasional diorama at a natural history museum, the marvelous diversity of West End illusion has been entirely extinguished. All those moving panoramas and magic lantern shows were wiped off the map by a single new technology: the cinema.

In a sense, the temples of illusion helped create the technology that ultimately destroyed them. Most of their innovations turned out to be dead ends: no one bothers to paint a thousand-foot canvas anymore, for obvious reasons. But the Phantasmagoria and the Panorama and their many peers did help solidify a new convention: that human beings would pay money to crowd together in a room and lose themselves in immersive, illuminated images. In 1820, this practice was limited to a tiny portion of the planet's population, clustered in a few affluent cities, but it would soon become a worldwide phenomenon with the rise of motion pictures.

The pattern that played out on the streets of the West End would recur many times over the subsequent decades. A cluster of innovations emerges, all experimenting with different variations on a single theme, until one specific solution arises that reaches critical mass and kills off its rivals. Think of the ecosystem of computer networks in the early 1990s: proprietary services like AOL and CompuServe; file-sharing protocols like Fetch or Gopher; private bulletin-board communities like The WELL or ECHO; hypertext experiments like Storyspace or HyperCard. Behind all these marginal new platforms, a shared consensus was visible: people were going to start consuming and sharing news, documents, personal information, and other media through hypertextual networks. But it was unclear whether a single platform would unite all these disparate activities, until the World Wide Web became the de facto standard in the mid-1990s. The process happened faster than it did in the days of West End illusion, but the underlying pattern was the same: early experiments, followed by explosive diversity, followed by radical consolidation.

The innovation that triumphs at the end of this sequence is often inferior in many ways to its rivals: remember that cinema, for all its advantages, lacked color for its first fifty years, and even in the age of 3-D IMAX, movies lack the 360-degree vista of the Panorama. But cinema was not a classic Clayton Christensen–style disruption where an inferior but cheap new product wipes out a more fully featured but expensive rival. As alluring as the mechanical dancers were at Merlin's, no one mistook them for genuine human beings. Once you could project images of actual people onto the screen—dancing and gesticulating and emoting, even without color or sound—the appeal of magic-lantern specters dissolved into thin air. To be sure, some of the most popular early films took their cues from the West End illusionists, most spectacularly in the special-effects-laden shorts of Georges Méliès. But one could argue that the key innovation that secured the dominance of cinema was not the camera or projector, nor the trick shots that Méliès pioneered, but rather the invention of the close-up. Early or "primitive" cinema—as the film scholars refer to it—was effectively a continuation of the West End spectacles: either immersing the viewer in the experience of a distant place, the way Banvard's moving panorama took the audience down the Mississippi, or subjecting the viewer to a series of mesmerizing special effects. Most of these films were shot in a way that mimicked the audience's perspective in a traditional theater: a single long shot with the actors framed by a stage set. But, starting in the 1910s, directors like D. W. Griffith began tinkering with the close-up, a technique that brought the spectator into a kind of intimate relationship with the actors that no stage production could achieve. That was the moment when cinema left the world of amusement and became art.

All of these elements—the proto-cinemas of the West End, the transformative power of moving images, the formal seduction of the close-up—make it clear just how hard it is to pinpoint exactly when the cinema itself was invented. Like most important technologies, cinema was an amalgam of very different innovations, which themselves drew upon varied forms of expertise and were developed on widely divergent time scales. The practice of projecting an image onto a screen by shining a stable light behind a semitransparent plate became commonplace with the rise of magic lanterns in the 1600s; the glass lenses used by both film cameras and projectors predate the magic lantern by a few centuries. Capturing images directly to a photosensitive material became possible in the early 1800s. The modern cinema that began to emerge in the second decade of the twentieth century drew upon centuries of innovations—in chemistry, optics, glassmaking, and mechanics—not to mention creative innovations like the tracking shot or the close-up. On top of all these breakthroughs, the cinema relied on a business model that had developed among the West End showmen: charging a fixed price for tickets to immerse oneself in a darkened chamber of illusion. That, too, was a kind of invention.

Motion pictures departed from the techniques of West End illusion in one key respect. In the Phantasmagoria or Banvard's moving panorama, the viewer's perception of motion was based on the actual movement of physical objects: the magic lantern rolling back and forth on its tracks, the steady unfurling of Banvard's thousand-yard painting. But the movement in moving pictures was merely a trick of the eye; to this day, every film or television show is constructed out of a stream of still images that our eye perceives as continuous motion. This phenomenon is commonly called persistence of vision, though there is intense debate in the scientific community over the neural mechanisms that make it possible. Like so many important elements of the human visual system, it was originally discovered by toymakers, in the spinning wheels and rotating drums of early nineteenth-century devices like the thaumatrope and the zoetrope, which spun through a dozen or so static images of a dancer or trotting horse, creating the illusion of movement. (*Thaumatrope* means wonder turner in Greek, while *zoetrop*e roughly translates to life turner, or wheel of life.) Whatever its biological roots, persistence of vision appears to be a universal property of the human eye: when still images are flashed at more than ten or twelve times a second, our eye stitches them together into a continuous flow.

On some basic level, this property of the human eye is a defect. When we watch movies, our eyes are empirically failing to give an accurate report of what is happening in front of them. They are seeing *something that isn't there.* Many technological innovations exploit the strengths that evolution has granted us: tools and utensils harness our manual dexterity and opposable thumbs; graphic interfaces draw on our powerful visual memory to navigate information space.

A crowd gathers to watch a magic lantern show

Two early thaumatropes, 1826

But moving pictures take the opposite approach: they succeed precisely because our eyes fail.

This flaw was not inevitable. Human eyesight might have just as easily evolved to perceive a succession of still images as exactly that: the world's fastest slide show. Or the eye might have just perceived them as a confusing blur. There is no evolutionary reason why the eye should create the illusion of movement at twelve frames per second; the ancestral environment where our visual systems evolved had no film projectors or LCD screens or thaumatropes. Persistence of vision is what Stephen Jay Gould famously called a spandrel—an accidental property that emerged as a consequence of other more direct adaptations. It is interesting to contemplate how the past two centuries would have played out had the human eye not possessed this strange defect. We might be living in a world with jet airplanes, atomic bombs, radio, satellites, and cell phones—but *without* television and movies. (Computers and computer networks would likely exist, but without some of the animated subtleties of modern graphical interfaces.) Imagine the twentieth century without propaganda films, Hollywood, sitcoms, the televised Nixon-Kennedy debate, the footage of civil rights protesters being fire-hosed, *Citizen Kane*, the Macintosh, James Dean, *Happy Days*, and *The Sopranos*. All those defining experiences exist, in part, because natural selection didn't find it necessary to perceive still images accurately at rates above twelve frames a second—and because hundreds of inventors, tinkering with the prototypes of cinema over the centuries, were smart enough to take that imperfection and turn it into art.

Art is the aftershock of technological plates shifting. Sometimes the aftershock is slow in arriving. It took the novel about three hundred years to evolve into its modern form after the invention of the printing press. The television equivalent of the novel—the complex serialized drama of *The Wire* or *Breaking Bad*—took as long as seventy years to develop, depending on where you date its origins. Sometimes the aftershocks roll in quickly: rock 'n' roll emerged almost instantaneously after the invention of the electric guitar. But some new artistic forms are deeply bound up in technological innovation: Brunelleschi using mirrors to trick his own eye into painting with linear perspective; Walter Murch inventing surround sound to capture the swirling chaos of Vietnam in *Apocalypse Now*. The artist's vision demands new tools to realize that vision, and every now and then the artist turns out to be a toolmaker as well. When those skills overlap in a single person, things move fast.

A Disney animator works on cels for the film Snow White

You can make the argument that the single most dramatic acceleration point in the history of illusion occurred between the years of 1928 and 1937, the years between the release of *Steamboat Willie*, Walt Disney's breakthrough sound cartoon introducing Mickey Mouse, and the completion of Disney's masterpiece, *Snow White*, the first long-form animated film in history. It is hard to think of another stretch where the formal possibilities of an artistic medium expanded in such a dramatic fashion, in such a short amount of time. *Steamboat Willie* is rightly celebrated for what it brought to the animator's art: synchronized sound and a memorable character with a defined personality. But it's also worth watching today for what it is conspicuously missing: there's no color, no spoken dialogue, only the hint of three-dimensionality. The story, only seven minutes long, revolves entirely around simple visual gags. *Steamboat Willie* was closer to a flip-book animation with a grainy soundtrack attached to liven things up. Viewed next to *Snow White*, *Willie* seems like it belongs to another era altogether, like comparing Méliès's *A Trip to the Moon* from 1902 with Orson Welles's *Citizen Kane*, made almost forty years later. Disney managed to compress a comparable advance in complexity into nine short years.

And even Welles, for all his genius, relied on innovations that had been pioneered by other filmmakers before him: Griffith's close-up; the sound synchronization introduced in *The Jazz Singer*, the dolly shot popularized by the Italian director Giovanni Pastrone. The great leap forward that Disney achieved with *Snow White* was propelled, almost exclusively, by imaginative breakthroughs inside the Disney studios. To produce his masterpiece, Disney and his team had to reinvent almost every tool that animators had hitherto used to create their illusions. The physics of early animation were laughably simplified; gravity played almost no role in the Felix the Cat or *Steamboat Willie* shorts that amused audiences in the 1920s. For *Snow White*, Disney wanted the entire animated world to play by the physical laws of the real world. Before *Snow White*, one animator recalled, "no one thought of clothing following through, sweeping out, and dropping a few frames later, which is what it does naturally. Disney commissioned thousands of slow-motion photographic studies that the animators could analyze to mimic the micro-behaviors of muscles, hair, smoke, glass breaking, birds flying, and countless other physical movements that had to be re-created with pen and ink. The team also inaugurated a drawing technique they called overlapping motion, in which all the characters were drawn engaging in constant, if subtle, physical activity, rather than just cycling through a series of static poses. These artistic innovations required a new way to test visual experiments before committing them to final print. Disney's team began sketching out ideas on cheap negative film that could be quickly processed and projected onto a tiny Moviola screen. They called these experimental trials pencil tests. The sheer length of *Snow White* required additional tools to map the overarching narrative; for that, the "storymen" on Disney's team hit upon the idea of taking sketches corresponding to each major scene and pinning them to a large corkboard that let Disney and his collaborators take in the narrative in a single glance, inaugurating the tradition of "storyboarding" that would become a ubiquitous practice in Hollywood, for both live-action and animated films.

Sound and color also forced Disney and his team to conjure up new solutions. Creating the illusion of spoken dialogue emerging from a human character's mouth required a level of synchronization and anatomical detail that early sound cartoons, like *Steamboat Willie*, had not required. While Disney had partnered with a new start-up called Technicolor to add a full palette to the final print of *Snow White*, the actual animation cels had to be painted in-house by the animation team. They ended up concocting a new kind of paint, using a gum arabic base that was "rewettable," enabling the animators to fix small problems without tossing out the entire cel. Disney even purchased a cutting-edge tool called a spectraphotometer to measure color levels precisely, given the challenge of converting them into the less accurate Technicolor format.

The most impressive technical breakthrough behind *Snow White* was the multiplane camera that Disney and his team built to create the signature sense of visual depth that *Snow White* introduced to animation. Before *Snow White*, animated films lived in a two-dimensional world, with only a hint of depth provided by an occasional linear perspective trick borrowed from Brunelleschi. But mostly they looked like a series of drawings on white paper that had somehow come to life. Most animations did use semitransparent character and background cels, layered on top of each other, so that animators wouldn't have to redraw the entire mise-en-scène for each frame. For *Snow White*, Disney hit upon the idea of multiple layers corresponding to different points in the virtual space of the movie, and separating those cels physically from each other while filming them: one layer for the characters in the foreground, one for a cottage behind them, another for the trees behind the cottage, and so on. By moving the position of the camera in tiny increments for each frame, a parallax effect could be simulated, creating an illusion of depth even more profound than the one Brunelleschi had invented five hundred years before. The multiplane camera was such an impressive feat of engineering that it warranted an extensive write-up in *Popular Science*:

> *[The device] consists of four vertical steel posts, each carrying a rack along which as many as eight carriages may be shifted both horizontally and vertically. On each carriage rides a frame containing a sheet of celluloid, on which is painted part of the action or background. Resembling a printing press, the camera stands eleven feet tall and is six feet square. Made with almost micrometer precision, it permits the photographing of foreground and background cels accurately, even when the first is held firmly in*

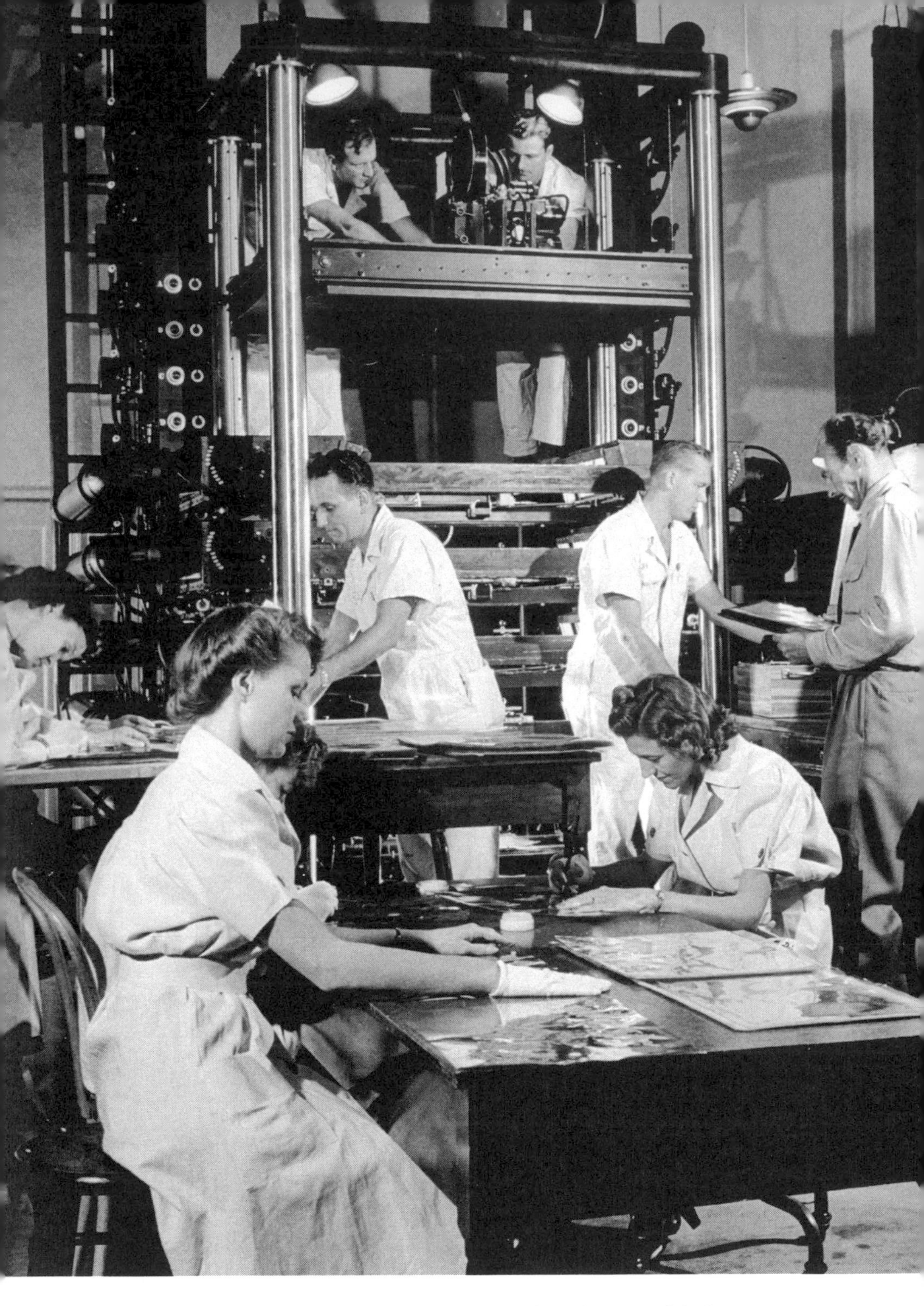

Disney studio cameramen shooting pictures with the multiplane camera as women ink in colors below

> *place two feet from the lens and the lowest rests in its frame nine feet away. Where the script calls for the camera to "truck up" for a close-up, the lens actually remains stationary, while the various cels are moved upward. By this means, houses, trees, the moon, and any other background features, retain their relative sizes.*

All of these technical and procedural breakthroughs summed up to an artistic one: *Snow White* was the first animated film to feature both visual and emotional depth. It pulled at the heartstrings in a way that even live-action films had failed to do. This, more than anything, is why *Snow White* marks a milestone in the history of illusion. "No animated cartoon had ever looked like *Snow White*," Disney's biographer Neil Gabler writes, "and certainly none had packed its emotional wallop." Before the film was shown to an audience, Disney and his team debated whether it might just be powerful enough to provoke tears—an implausible proposition given the shallow physical comedy that had governed every animated film to date. But when *Snow White* debuted at the Carthay Circle Theatre, near L.A.'s Hancock Park, on December 21, 1937, the celebrity audience was heard audibly sobbing during the final sequences where the dwarfs discover their poisoned princess and lay garlands of flowers on her. It was an experience that would be repeated a billion times over the decades to follow, but it happened there at the Carthay Circle first: a group of human beings gathered in a room and were moved to tears by hand-drawn static images flickering in the light.

In just nine years, Disney and his team had transformed a quaint illusion—the dancing mouse is whistling!—into an expressive form so vivid and realistic that it could bring people to tears. Disney and his team had created the ultimate illusion: fictional characters created by hand, etched onto celluloid, and projected at twenty-four frames per second, that were somehow so believably human that it was almost impossible *not* to feel empathy for them.

Those weeping spectators at the *Snow White* premiere signaled a fundamental change in the relationship between human beings and the illusions concocted to amuse them. Complexity theorists have a term for this kind of change in physical systems: *phase transitions.* Alter one property of a system—lowering the temperature of a cloud of steam, for instance—and for a while the changes are linear: the steam gets steadily cooler. But then, at a certain threshold point, a fundamental shift happens: below 212 degrees Fahrenheit, the gas becomes liquid water. That moment marks the phase transition: not just cooler steam, but something altogether different. When you cross the boundary of the phase transition, new possibilities emerge, ones you might have never imagined while contemplating the earlier state. In the case of liquid water, one of those new possibilities was life itself.

Twelve frames per second is the perceptual equivalent of the boundary between gas and liquid. When we crossed that boundary, something fundamentally different emerged: still images came to life. The power of twelve frames per second was so irresistible that it even worked with hand-drawn characters pulled from a storybook. But like the phase transitions of water, passing that threshold—and augmenting it with synchronized sound—unleashed other effects that were almost impossible to predict in advance. The consumers of illusion at the beginning of the nineteenth century wouldn't be at all surprised to find that people two centuries later were gathering in dark rooms to be startled and surprised by special effects. But they *would* be surprised by something else in the culture: the enormous emotional investment that people have in the lives of other people they have never met, people who have done almost nothing of interest other than appear on a screen. The phase transition of twelve frames per second introduced a class of people virtually unknown until the twentieth century: celebrities.

Fame, of course, is an old story, as old as history itself. Kings, military heroes, statesmen, clerics, prophets—all possessed lives that reverberated far beyond their circles of immediate acquaintance. As the historian Fred Inglis describes it, "celebrity was inseparable from the public acknowledgement of achievement." We can trace the origins of modern celebrity culture back to the coffeehouse chatter of early publications like the *Tatler* that shared moderately suggestive stories of London's aristocracy in the early eighteenth century. Those voices were amplified by the end of the century with salacious tales about the debauched life of the Prince Regent; shortly thereafter, Lord Byron established a template for artistic genius and renegade sexual adventurism that would be emulated by a thousand rock stars in the postwar years. Stars of the stage like Sarah Siddons or Sarah Bernhardt generated intrigue about their private lives before the first Hollywood gossip columns appeared in the 1940s. But however much the prurient interest of the general public in these figures may remind us of modern celebrity culture, one key difference remains: the princes and poets and actresses that garnered attention in the age before television and cinema were living genuinely extraordinary lives—thanks either to the extreme good fortune of being born into a royal family or to their own achievements as artists or writers or actors. The surplus fixation of fame was still grounded in the use-value of an exceptional life. Today, of course, that pool of celebrity has widened dramatically, a phenomenon famously captured by Andy Warhol in his fifteen-minutes-of-fame witticism, but also by Daniel Boorstin in his 1961 classic, *The Image*: "We still try to make our celebrities stand in for the heroes we no longer have, or for those who have been pushed out of our view. We forget that celebrities are known primarily for their well-knownness." Page through the gossip magazines of today and you will be astounded to see how even the Hollywood celebrities have taken a backseat to the stars of reality television. Barrels of ink are spilled each day sharing breathless accounts of date nights and pregnancy bumps in the lives of people who appeared on *The Bachelor* five years ago. Celebrity was once earned through an extraordinary career; later, it could be achieved by *pretending* to be extraordinary people onstage or on the screen. But today celebrity is just as likely to belong to people who have no claim to fame other than the fact that their ordinary lives appear on television.

It is hard not to feel that these shows—and the long echo of gossip that trails behind them—are simply a colossal waste of time. But as banal as these new "personalities" are, their existence still suggests an interesting question: Why did this kind of celebrity culture only emerge in recent decades? The answer, I think, comes down yet again to the power of illusion, its ability to distort our perception of reality, making it impossible for us to *not* see things that are, empirically, not there. At twelve frames per second, with synchronized sound and close-ups, it is almost impossible for human beings *not* to form emotional connections with the people on-screen. (Disney made it clear that you didn't even need actual people!) We naturally feel interest in the everyday ups and downs of our close friends and family. Twelve frames a second tricks the brain into feeling that same level of intimacy with people we will never meet in person, what Inglis calls "knowability combined with distance." When the tinkerers of the 1830s were exploiting persistence of vision to make a horse come to life in the circular motion of the thaumatrope, it never occurred to them that the perceptual error they were exploiting would one day cause people to weep and bristle at the mundane actions of total strangers living thousands of miles from them. But that is the strange cognitive alchemy that twelve frames per second helped stir into being. Persistence of vision is an evolutionary accident that created the conditions of possibility for a cultural accident. The modern celebrity is a spandrel of a spandrel.

It is possible—maybe even likely—that a further twist awaits us. Recall the "irresistible eyes" of the mechanical dancer that so entranced Charles Babbage in Merlin's attic. Those robotic facial expressions would seem laughable to a modern viewer, but animatronics has made a great deal of progress since then. There may well be a comparable threshold in simulated emotion—via robotics or digital animation—that makes it near impossible for humans not to form emotional bonds with a simulated being. We knew the dwarfs in *Snow White* were not real, but we couldn't keep ourselves from weeping for their lost princess in sympathy with them. Imagine a world populated by machines or digital simulations that fill our lives with comparable illusion, only this time the virtual beings are not following a storyboard sketched out in Disney's studios, but instead responding to the twists and turns and unmet emotional needs of our own lives. (The brilliant Spike Jonze film *Her* imagined this scenario using only a voice, though admittedly the voice belonged to Scarlett Johansson.) There is likely to be the equivalent of a Turing Test for artificial *emotional* intelligence: a machine real enough to elicit an emotional attachment. It may well be that the first simulated intelligence to trigger that connection will be some kind of voice-only assistant, a descendant of software like Alexa or Siri—only these assistants will have such fluid conversational skills and growing knowledge of our own individual needs and habits that we will find ourselves compelled to think of them as more than machines, just as we were compelled to think of those first movie stars as more than just flickering lights on a fabric screen. Once we pass that threshold, a bizarre new world may open up, a world where our lives are accompanied by simulated friends. In a strange way, these virtual companions might be more authentic than the simulated friends of reality TV; at least the robots and virtual humans would acknowledge your existence and engage directly with your shifting emotional states, unlike the Kardashians. The ghost makers and automaton designers of the eighteenth century first tapped the power of illusion to terrify or amuse us; their descendants in the twenty-first century may draw on the same tools to conjure up other feelings: empathy, companionship, even love.

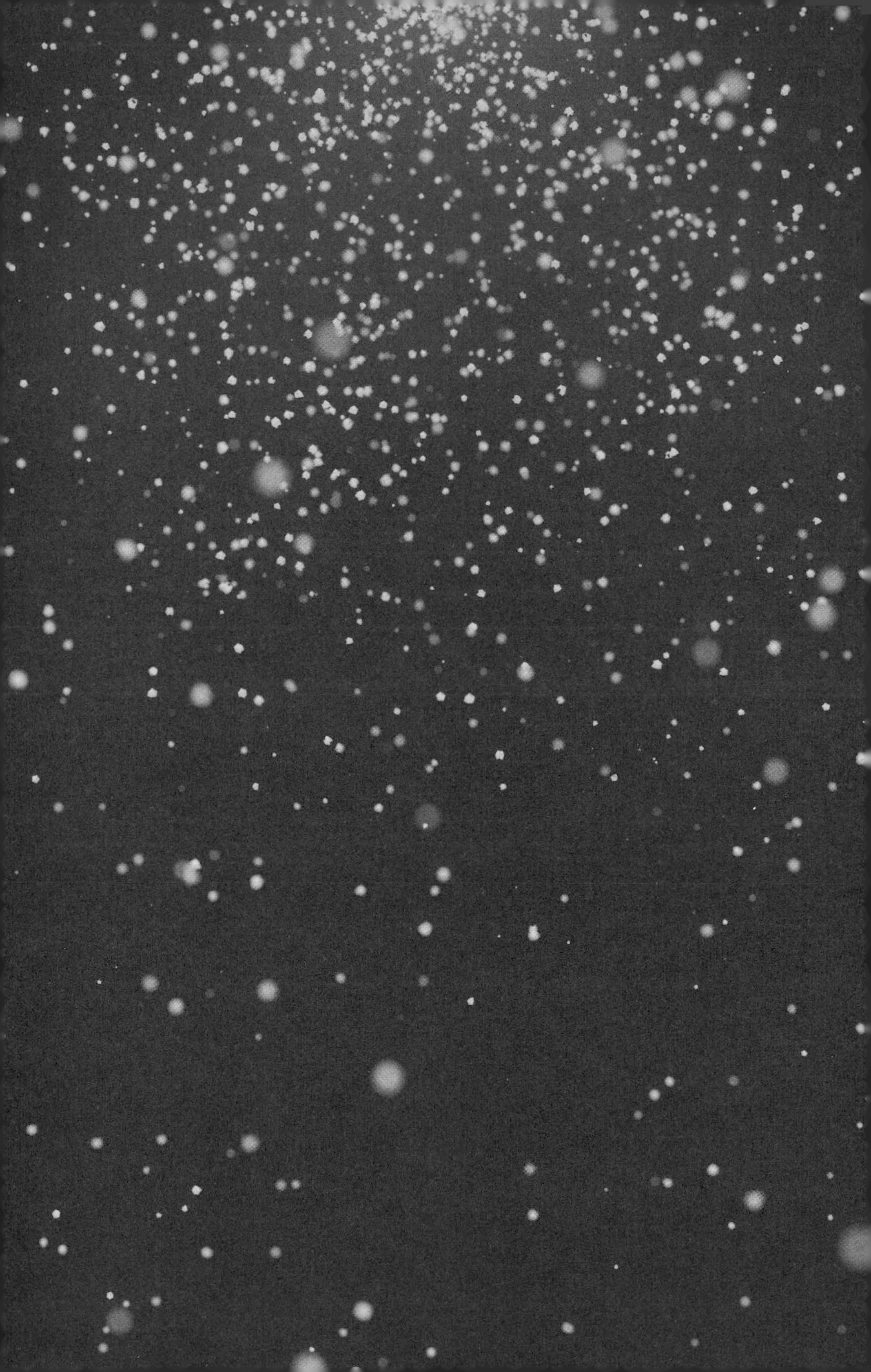

5

Games

The Landlord's Game

In the second half of the thirteenth century, a Dominican friar from the Lombard region of modern-day Italy began delivering a series of sermons that meditated on the proper roles of different social groups: royalty and merchants, clergy and farmers. Over the years, his name has been warped into an almost comical number of variations, including Cesolis, Cessole, Cesulis, Cezoli, de Cezolis, de Cossoles. But historians now generally refer to him as Jacobus de Cessolis. We know next to nothing about Cessolis, only that his sermons were well received by his monastic brothers, and the friar was encouraged to translate them to the page.

This was still more than a century and a half before Gutenberg, so the friar's words were inscribed by hand, and not mass-printed when they first began to circulate. Yet Cessolis's text would end up playing a crucial role in the early history of the printing press. Something about his message resonated with the popular mood, and in the decades that followed its first appearance, hand-copied versions of Cessolis's original manuscript swept across Europe. "The original was not just copied," the historian H. L. Williams notes. "It was translated, abbreviated, modified, turned into verse, and otherwise edited by substitutions and emendations that made the material relevant to other regions and cultures . . . Some versions grew to more than four times the length of Cessolis' original." By the time Gutenberg introduced his movable-type technology, Cessolis's manuscript was, by some accounts, surpassed only by the Bible in its popularity. More than fifteen separate editions, in almost as many languages, were published in the first decades after Gutenberg originally printed his Bible. It even caught the eye of the entrepreneurial British printer William Caxton, who translated and published it out of a printing shop in Bruges. It was the second book ever printed in the English language.

What was the title of this manifesto? Cessolis gave it two. The first was more formal: *The Book of the Manners of Men and the Offices of the Nobility.* But the second had an unusual twist. He called it *The Game of Chess.*

The Game of Chess is one of those books that would be impossible to place on a modern bookstore shelf. It defies sensible genre categorization. It was, in part, a straightforward guide to the rules of chess, with sections devoted to each of the pieces and the rules that govern their behavior. But out of the game's elemental rules, Cessolis constructed an elaborate allegorical message: the king's behavior on the board reflected the proper behavior of real-world kings; the knight's reflected the proper behavior of real-world knights. (To extend his societal analysis, Cessolis made each pawn represent a specific group of commoners: farmers, blacksmiths, money-changers, and so on.) It was a baffling hybrid: a profound sociological treatise bound together with a game guide. It was as though Studs Terkel had bundled his 1970s classic *Working* with a cheat sheet for Ms. Pac-Man. The combination makes for some entertaining leaps, where Cessolis jumps from the minutiae of chess strategy to broad generalizations about medieval lifestyles without the slightest pause:

> *Both knights have three possible openings. The white knight may open toward the right onto the black square in front of the farmer. This is reasonable, because the farmer works the fields, tills and cultivates them. The knight and his horse receive food and nourishment from the farmer and protect him in return. The second opening is to the black square in front of the clothier. This move is also reasonable since the knight ought to protect the man who makes his clothes. The third opening is toward the left onto the black square in front of the king, but only if the merchant is not there. This is reasonable since the knight is supposed to guard and protect the king as he would himself.*

Some of the friar's lessons were sensible enough: "The king has dominion, leadership, and rank over everyone, and for that reason should not travel much outside his kingdom." Others seem less advisable to a modern reader, because the rules of society *and* the rules of chess have changed: "If women want to remain chaste and pure they should not sit by the door to the gardens, they should avoid venturing into the streets, and they should not forget all their ladylike manners." Avoiding "venturing out into the streets" might have been a good strategy for the queen in medieval games of chess, but that is only because that piece wouldn't take on its extensive powers until the 1500s; in Cessolis's time, the queen was severely limited in its mobility on the board.

But as amusing as it is to chuckle over the eccentricities of *The Game of Chess*, it is more instructive to contemplate the book's cultural significance. Perhaps unwittingly, the friar was dismantling a vision of social organization that had been dominant for at least a thousand years: the *body* politic, the image of society as a single organism, directed—inevitably—by the metaphoric "head of state." Instead, Cessolis proposed a different model: independent groups, governed by contractual and civic obligations, but far more autonomous than the "body politic" metaphor had implied. This amounted to a profound shift in social consciousness. For more than a thousand years, social order had been imagined in physiological terms. "If the head of a body decided that the body should walk," the historian Jenny Adams writes, "the feet would have to follow." But the chessboard suggested another way of connecting social pieces: a society regulated by laws and contracts and conventions. "The chess allegory imagines its subjects to possess independent bodies in the form of pieces bound to the state by rules rather than biology," Adams explains. "If the chess king advances, the pawns are not beholden to do the same."

The social transformation mapped out on Cessolis's chessboard would eventually ripple across the map of Europe itself, with the rise of the Renaissance guild system and a wave of legal codes that endowed merchants and artisans with newfound freedoms. Social positions were not defined as God-given inevitabilities, but rather emerged out of legal and ethical conventions. Authority no longer trickled down from above, but instead was established through a web of consensual relationships, in many cases defined by contracts. This was not yet a radical secularization of society; kings and bishops still had important roles. But it was a crucial early step that led toward the truly secular state that would arise five hundred years later: a society governed ultimately by laws and not monarchs.

How big a role did *The Game of Chess* have in triggering this shift? It is difficult to say for certain. It may well be that Cessolis was simply popularizing a conceptual shift that was already under way, and thus the book's success lay in the way it *explained* a historic change that must have been bewildering for the citizens trying to make sense of it. Or perhaps the book's popularity helped hasten the transformation itself, as the chess allegory encouraged readers to seek

A miniature of Otto IV of Brandenburg playing chess with a lady, circa 14th century

out the kind of contractual autonomy the book celebrated. Either way, it is clear that the game itself provided a new way of thinking about society, a new framework that could be used to understand some of the most important issues in civic life. This turns out to be one of the key ways in which the seemingly frivolous world of game play affects the "straight" world of governance, law, and social relations. The experimental tinkering of games—a parallel universe where rules and conventions are constantly being reinvented—creates a new supply of metaphors that can then be mapped onto more serious matters. (Think how reliant everyday speech is on metaphors generated from games: we "raise the stakes"; we "advance the ball"; we worry about "wild cards"; and so on.) Every now and then, one of those metaphors turns out to be uniquely suited to a new situation that requires a new conceptual framework, a new way of imagining. A top-down state could be described as a body or a building—with heads or cornerstones—but a state governed by contractual interdependence needed a different kind of metaphor to make it intelligible. The runaway success of *The Game of Chess*—first as a sermon, then as a manuscript, and finally as a book—suggests just how valuable that metaphor turned out to be.

We commonly think of chess as the most intellectual of games, but in a way its greatest claim to fame may be its allegorical power. No other game in human history has generated such a diverse array of metaphors. "Kings cajoled and threatened with it; philosophers told stories with it; poets analogized with it; moralists preached with it," writes chess historian David Shenk. "Its origins are wrapped up in some of the earliest discussions of fate versus free will. It sparked and settled feuds, facilitated and sabotaged romances, and fertilized literature from Dante to Nabokov." Metaphoric thinking can be mind-opening, the way Cessolis's allegory helped medieval Europe understand a new civic order. But it can also be restrictive, as the metaphor keeps you from seeing other possibilities that don't quite fit the mold. Sometimes it can be a bit of both. In the middle of the twentieth century, chess became a kind of shorthand way of thinking about intelligence itself, both in the functioning of the human brain and in the emerging field of computer science that aimed to mimic that intelligence in digital machines.

The very roots of the modern investigation into artificial intelligence are grounded in the game of chess. "Can [a] machine play chess?" Alan Turing famously asked in a groundbreaking 1946 paper. "It could fairly easily be made to play a rather bad game. It would be bad because chess requires intelligence . . . There are indications however that it is possible to make the machine display intelligence at the risk of its making occasional serious mistakes . . . What we want is a machine that can learn from experience . . . [the] possibility of letting the machine alter its own instructions." Turing's speculations form a kind of origin point for two parallel paths that would run through the rest of the century: building intelligence into computers by teaching them to play chess, and studying humans playing chess as a way of understanding our own intelligence. Those interpretative paths would lead to some extraordinary breakthroughs: from the early work on cybernetics and game theory from people like Claude Shannon and John von Neumann, to machines like IBM's Deep Blue that could defeat grandmasters with ease. In cognitive science, the litany of insights that derived from the study of chess could almost fill an entire textbook, insights that have helped us understand the human capacity for problem solving, pattern recognition, visual memory, and the crucial skill that scientists call, somewhat awkwardly, chunking, which involves grouping a collection of ideas or facts into a single "chunk" so that they can be processed and remembered as a unit. (A chess player's ability to recognize and often name a familiar sequence of moves is a classic example of mental chunking.) Some cognitive scientists compared the impact of chess on their field to *Drosophila*, the fruit fly that played such a central role in early genetics research.

But the prominence of chess in the first fifty years of both cognitive and computer science also produced a distorted vision of intelligence itself. It helped cement the brain-as-computer metaphor: a machine driven by logic and pattern recognition, governed by elemental rules that could be decoded with enough scrutiny. In a way, it was a kind of false tautology: brains used their intelligence to play chess; thus, if computers could be taught to play chess, they would be intelligent. But human intelligence turns out to be a much more complex beast than chess playing suggests. When Deep Blue finally defeated world chess champion Gary Kasparov in 1997, it marked a milestone for computer science, but Deep Blue itself was largely helpless if you wanted to ask it about anything other than chess. It would have captured Cessolis's king in a matter of minutes, but if you wanted to know something about *actual* kings, Cessolis would be infinitely more informative. (Noam Chomsky famously declared that Deep Blue's victory was about as interesting as a bulldozer winning the Olympics for weight lifting.) Today, the study of intelligence has greatly diversified: we know that the skills necessary to play chess are only a small part of what it means to be smart.

Cessolis was hardly alone in conceiving the game board as a platform for moral instruction. Up until the end of the nineteenth century, most American board games were explicitly designed to impart ethical or practical lessons to their players. One popular game of the 1840s, the Mansion of Happiness, embedded a stern Puritan worldview in its game play. Many passages from the game's rule book sound more like a Cotton Mather sermon than an idle amusement:

> *WHOEVER possesses PIETY, HONESTY, TEMPERANCE, GRATITUDE, PRUDENCE, TRUTH, CHASTITY, SINCERITY . . . is entitled to Advance six numbers toward the Mansion of Happiness. WHOEVER gets into a PASSION must be taken to the water and have a ducking to cool him . . .*

WHOEVER posses[ses] AUDACITY, CRUELTY, IMMODESTY, or INGRATITUDE, must return to his former situation till his turn comes to spin again, and not even think of HAPPINESS, much less partake of it.

Milton Bradley launched his gaming empire in 1860 with the Checkered Game of Life, the distant ancestor of the modern Game of Life. (Both games use a spinner instead of dice, a randomizing device introduced by Bradley to differentiate his game from the long-standing association between dice games and gambling.) As gaming historian Mary Pilon notes, Bradley's creation was quite a bit darker than its contemporary version: "[The] board had an Intemperance space that led to Poverty, a Government Contract space that led to Wealth, and a Gambling space that led to Ruin. A square labeled Suicide had an image of a man hanging from a tree, and other squares were labeled Perseverance, School, Ambition, Idleness, and Fat Office."

But the most intriguing—and tragically overlooked—heir to Cessolis was a woman named Lizzie Magie who managed to conceive of the game board as a vehicle not of conventional religious instruction, but rather social and political revolution. Born in Illinois in 1866, Magie had an eclectic and ambitious career even by suffragette standards. She worked at various points as a stenographer, poet, and journalist. She invented a device that made typewriters more efficient, and filed for a patent for it in 1893. She worked part-time as an actress on the stage. For a long time, her greatest claim to fame came through an act of political performance art, placing a mock advertisement in a local paper that put herself on the market as a "young woman American slave"—protesting the oppressive wage gap between male and female salaries, and mocking the mercenary nature of many traditional marriages. Magie was also a devotee of the then-influential economist Henry George, who had argued in his 1879 best-selling book *Progress and Poverty* for an annual "land-value tax" on all land held as private property—high enough to obviate the need for other taxes on income or production. Many progressive thinkers and activists of the period integrated "Georgist" proposals for single-tax plans into their political platforms and stump speeches. But only Lizzie Magie appears to have decided that radical tax reform might make compelling subject matter for a board game.

Magie began sketching the outlines of a game she would come to call the Landlord's Game. In 1904, she published a brief outline of the game in a Georgist journal called *Land and Freedom*. Her description would be immediately familiar to most grade schoolers in dozens of countries around the world:

> *Representative money, deeds, mortgages, notes and charters are used in the game; lots are bought and sold; rents are collected; money is borrowed (either from the bank or from individuals), and interest and taxes are paid. The railroad is also represented, and those who make use of it are obliged to pay their fare, unless they are fortunate enough to possess a pass, which, in the game, means throwing a double. There are two franchises: the water and the lighting; and the first player whose throw brings him upon one of these receives a charter giving him the privilege of taxing all others who must use his light and water. There are two tracts of land on the board that are held out of use—are neither for rent nor for sale—and on each of these appear the forbidding sign: "No Trespassing. Go to Jail."*

Magie had created the primordial Monopoly, a pastime that would eventually be packaged into the most lucrative board game of the modern era, though Magie's role in its invention would be almost entirely written out of the historical record. Ironically, the game that became an emblem of sporty capitalist competition was originally designed as a critique of unfettered market economics. Magie's version actually had two variations of game play, one in

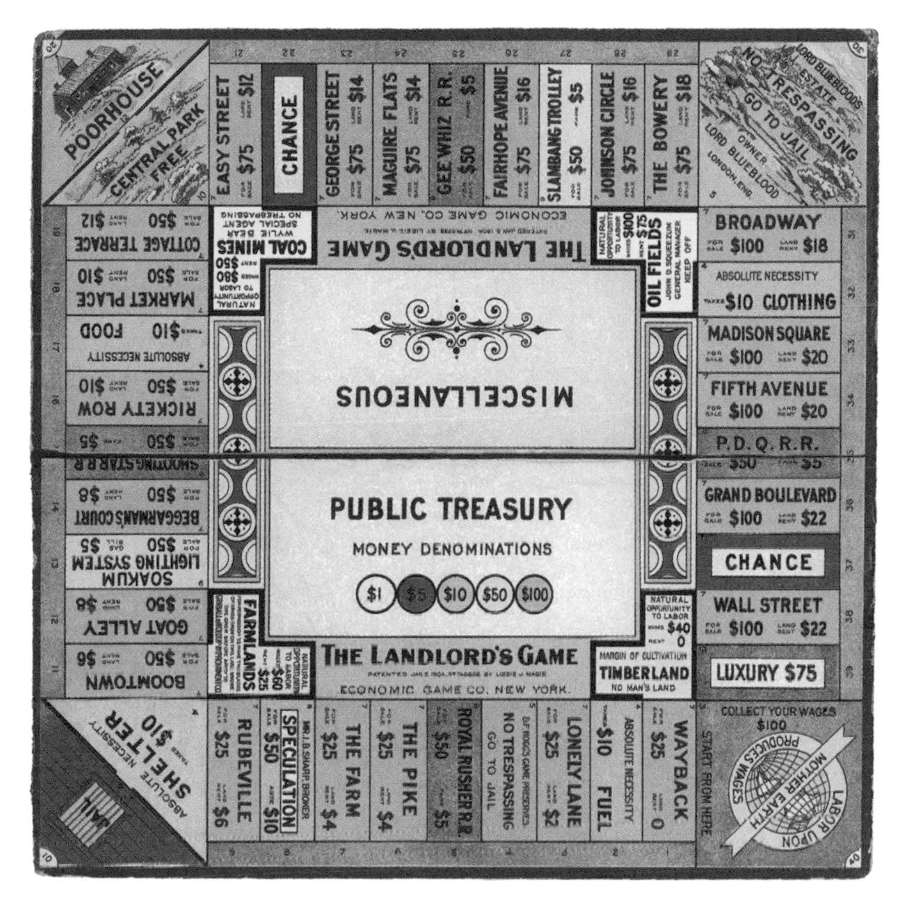

The playing board for The Landlord's Game, Monopoly's predecessor

which players competed to capture as much real estate and cash as possible, as in the official Monopoly, and one in which the point of the game was to share the wealth as equitably as possible. (The latter rule set died out over time—perhaps confirming the old cliché that it is simply less fun to be a socialist.) Either way you played it, however, the agenda was the same: teaching children how modern capitalism worked, warts and all. In its way, The Landlord's Game was every bit as moralizing as the Mansion of Happiness. "The little landlords take a general delight in demanding the payment of their rent," Magie wrote.

> *They learn that the quickest way to accumulate wealth and gain power is to get all the land they can in the best localities and hold on to it. There are those who argue that it may be a dangerous thing to teach children how they may thus get the advantage of their fellows, but let me tell you there are no fairer-minded beings in the world than our own little American children. Watch them in their play and see how quick they are, should any one of their number attempt to cheat or take undue advantage of another, to cry, "No fair!" And who has not heard almost every little girl say, "I won't play if you don't play fair." Let the children once see clearly the gross injustice of our present land system and when they grow up, if they are allowed to develop naturally, the evil will soon be remedied.*

The Landlord's Game never became a mass hit, but over the years it developed an underground following. It circulated, samizdat-style, through a number of communities, with individually crafted game boards and rule books dutifully transcribed by hand. Students at Harvard, Columbia, and the Wharton School played the game late into the night; Upton Sinclair was introduced to the game in a Delaware planned community called Arden; a cluster of Quakers in Atlantic City, New Jersey, adopted it as a regular pastime. As it traveled, the rules and terminology evolved. Fixed prices were added to each of the properties. The Wharton players first began calling it "the monopoly game." And the Quakers added the street names from Atlantic City that would become iconic, from Baltic to Boardwalk.

It was among that Quaker community in Atlantic City that the game was first introduced to a down-on-his-luck salesman named Charles Darrow, who was visiting friends on a trip from his nearby home in Philadelphia. Darrow would eventually be immortalized as the sole "inventor" of Monopoly, though in actuality he turned out to be one of the great charlatans in gaming history. Without altering the rules in any meaningful way, Darrow redesigned the board with the help of an illustrator named Franklin Alexander, and struck deals to sell it through the Wanamaker's department store in Philadelphia and through FAO Schwarz. Before long, Darrow had sold the game to Parker Brothers in a deal with would make him a multimillionaire. For decades, the story of Darrow's rags-to-riches ingenuity was inscribed in the Parker Brothers rule book: "1934, Charles B. Darrow of Germantown, Pennsylvania, presented a game called MONOPOLY to the executives of Parker Brothers. Mr. Darrow, like many other Americans, was unemployed at the time and often played this game to amuse himself and pass the time. It was the game's exciting promise of fame and fortune that prompted Darrow to initially produce this game on his own."

Both the game itself—and the story of its origins—had entirely inverted the original progressive agenda of Lizzie Magie's landlord game. A lesson in the abuses of capitalist ambition had been transformed into a celebration of the entrepreneurial spirit, its collectively authored rules reimagined as the work of a lone genius.

Game mythologies habitually seek out heroic inventors, even when those invention stories are on the very edges of plausibility. Every year, millions of baseball fans descend on the small town of Cooperstown, New York, to visit the Baseball Hall of Fame because Abner Doubleday invented the game in a cow pasture there in 1839. Most such origin stories overstate the role of the inventor, reducing a complex lineage down to a single individual genius. But in the case of Doubleday, the story appears to be an almost complete fabrication. An acclaimed Civil War general who had fired the first shot at Fort Sumter, Doubleday was pegged as baseball's founding father in 1907 by a commission headed by then National League president Abraham Mills. The commission had been assembled ostensibly to uncover the origins of the sport, though its unspoken objective was to prove that the roots of the American national pastime were, in fact, American. (Many believed, probably correctly, that the game evolved out of the British sport rounders.) And so the Mills Commission somewhat randomly declared that baseball's immaculate conception had occurred in Cooperstown in 1839, produced *sui generis* from the mind of Abner Doubleday. This origin story had all the right elements—military hero invents revered game in a flash of inspiration—except for the troubling fact that Doubleday seems to have had nothing to do with baseball whatsoever. In his voluminous collection of letters, he never once mentioned the game. Even worse, Doubleday was enrolled and living at West Point as a young cadet in 1839, his family having moved from Cooperstown the year before.

Invention stories like that of Abner Doubleday and the cow pasture so often fall apart with games because almost without exception our most cherished games have been the product of collective invention, usually involving collaborations that span national borders. Baseball, for instance, turns out to have a complicated lineage that includes rounders and cricket and an earlier game called stoolball. Citizens of Britain, Ireland, France, and the Netherlands, as well as the United States, played a role in the evolution of the game, although it is almost impossible to map the exact evolutionary tree. The conventional fifty-two-card deck that we now use for poker and solitaire evolved over roughly five hundred years, with important contributions coming from Egypt, France, Germany, and the United States. European football has roots that date back to Ancient Greece.

Chess, too, emerged out of multinational networks. Cessolis's *Game of Chess* managed to inspire so many different translations in part because the book's message resonated with broader developments in European society. But it also traveled well because *chess itself* had traveled well. Though the game had originated around 500 CE, by Cessolis's time, it was played throughout Europe, the Middle

East, and Asia. Today, we take it for granted that blockbuster movies play in theaters all across the globe, but a thousand years ago, before the age of exploration, cross-cultural exchange was severely limited by geographic and linguistic barriers. Chess was one of the first truly global cultural experiences. By medieval times, with players on three continents and in dozens of countries, even Christianity's geographic footprint looked small beside the long stride of chess.

That global expansion set a template that would be repeated hundreds of times by other games in the ensuing centuries. We may not take our games as seriously as we do our forms of governance or our legal codes or our literary novels, but for some reason games have a wonderful ability to cross borders. Games were cosmopolitan many centuries before the word entered the English vocabulary. And unlike religious or military encroachments, when games cross borders, they almost inevitably tighten the bonds between different nations rather than introducing conflict. European football is played professionally in almost every nation on the face of the earth. The global reach of games is even more pronounced in virtual gameplay. Consider the epic success of Minecraft, an immense online universe populated by players logging in from around the world. In the case of Minecraft, of course, the world of the game itself—and the rules that govern it—are being created by that multinational community of players, in the form of mods and servers programmed and hosted by Minecraft fans. McLuhan coined the term "global village" as a metaphor for the electronic age, but if you watch a grade-schooler constructing a virtual town in Minecraft with the help of players from around the world, the phrase starts to sound more literal.

The migratory history of chess, like that of most games, did not begin with some immaculate conception in the mind of some original genius game designer. As chess traveled across borders, new players in new cultures experimented with the rules. "Like the Bible and the Internet," Shenk writes, "[chess was] the result of years of tinkering by a large, decentralized group, a slow achievement of collective intelligence." Evolving out of an earlier Indian game called *chaturanga*, the first game that modern eyes would recognize as chess was played in Persia during the fifth century CE, a game called *chatrang*. The key ingredients were the same: a board divided into sixty-four squares, with two opposing armies of sixteen pieces each. The iconography drew on Indian culture: today's bishop, for instance, was an elephant on the *chatrang* board. The rules governing the movement of pieces also differed from today's game: the elephant could only move two squares diagonally, and the queen—known then as the minister—was as restricted as the king in its movement. The practice of announcing the king's imminent capture dates from this period: the odd phrase *checkmate* derives from the Persian words for king and defeat, *shah* and *mat*. Alternating black and white squares were introduced sometime around the game's first appearance in Europe, a few hundred years before Cessolis wrote his sermons. Shortly thereafter, regional versions of the game adopted a rule whereby pawns could be moved two squares in their opening move. In a shift allegedly inspired by the formidable Queen Isabella of Spain, herself an avid chess player, the queen became the most powerful piece on the board by the end of the fifteenth century.

Today, we take it for granted that software projects like Linux might be collectively authored by thousands of people scattered around the world, each contributing ideas to the project without any official affiliation or traditional management structure. This kind of global creation was almost unheard-of a thousand years ago: the limitations of transportation networks made invention and production largely local affairs. Games themselves—not the physical manifestation of the game, but the underlying *rules*—were among the first key cultural dishes to be cooked up in the global melting pot. (In a way, the closest equivalent to chess's cosmopolitan evolution are the scientific insights that followed a similar geographic path, from the Islamic Renaissance through medieval monasteries to the European

Enlightenment, with small but crucial additions and corrections added with each step of the journey.) Once again, a seemingly frivolous custom turns out to be an augur of future developments: if you were an aspiring futurologist in Cessolis's day and you were looking for clues about the future of invention and commerce—perhaps even a future where virtual encyclopedias would be written and edited by millions of people around the world—a good place to start would be by studying the games people were playing for fun, and the evolution of the rules that governed those games.

Any evolutionary tree contains branches that stop growing, promising new forms that die out. There are cultural extinctions just as there are biological ones. For every chess innovation that ultimately survived to the present day, there are others that perished. As chess found its way into Europe during the Middle Ages, some new players found the game too slow. (The queen's limited powers made it much more difficult to reach the endgame.) In part to speed things up, a new element was added to the game, one that would be anathema to the modern chess player: dice. "The wearingness which players experienced from the long duration of the game when played right through [is the reason] dice have been brought into chess, so that it can be played more quickly," a Spanish player wrote in 1283. A roll of the dice determined which pieces could be moved, bringing chess, briefly, from the realm of pure logic and perfect information into the realm of chance.

For whatever reason—perhaps because the introduction of randomness undermined precisely what made chess such a fascinating game—the dice-based version of chess did not survive. Chess didn't last long as a game of chance. But for all their association with gambling and lesser intellectual pursuits, games of chance would end up transforming society in ways that arguably exceed the impact of chess. And at the center of that revolution was the material design of the dice themselves.

Girolamo Cardano

In 1526, a young medical student from Padua named Girolamo Cardano found himself losing at cards in a Venetian gambling house. Faced with an alarming financial loss, Cardano determined that the deck of cards had been marked, and the game had been rigged against him from the start. Enraged at this deception, the twenty-five-year-old Cardano slashed his opponent in the face with a dagger, grabbed his lost money, and ran out into the streets of Venice, where he promptly fell into one of the canals. This episode seems to be a representative one for Cardano, who managed to live a spectacularly interesting—if somewhat debauched—life. He was apparently "hot tempered, single-minded, and given to women . . . cunning, crafty, sarcastic, diligent, impertinent, sad and treacherous, miserable, hateful, lascivious, obscene, lying, obsequious . . ." (This parade of adjectives, mostly critical, comes from Cardano himself, who wrote them in his own salacious autobiography.) For several years in his youth, his primary source of income came through gambling with cards and dice.

Cardano was also, as it happens, a gifted mathematician, and at some point in his early sixties he decided to combine his interests in a book called *Liber de Ludo Aleae*, or *The Book of Games of Chance*. Like Cessolis's work before him, Cardano's book was in part a game guide, a manual of sorts for the aspiring gambler, helping to inaugurate a somewhat less erudite branch of the publishing world that would eventually take the shape of countless how-to books for beating the odds in Las Vegas. But to make his advice actually useful, Cardano had to do something that most of his successors didn't have to bother with: he had to invent an entirely new field of mathematics.

Games of chance are among the oldest cultural artifacts known to man. The pharaohs of Egypt played dicelike games with what were called *astragali*, fashioned out of the anklebones of animals. A game resembling modern backgammon was discovered in the Royal

An early version of backgammon discovered in the Royal Tomb of Ur in Southern Iraq, dating back to 2600 BCE

Tomb of Ur, dating back to 2600 BC. Both the Greeks and Romans played obsessively with *astragali*. There seems to be a deep-seated human interest in chance and randomness, manifest in these intricate game pieces that have survived millennia, still recognizable to the modern eye. To play a game of chance is, in a sense, to rehearse for the randomness that everyday life presents, particularly in a prescientific world where basic circuits and patterns in nature had not yet been perceived.

But there are patterns, too, in the random outcomes of games of chance. Roll two modern dice and the number on each individual die will be random, but the sum of the two numbers will be more predictable: a seven is slightly more likely than an eight, and far more likely than a twelve. The concept is simple enough that young children can grasp it if it is explained to them: there are more ways of making a seven with two dice, and so the seven has a higher probability of being the result. And yet, despite the simplicity of the concept, it appears to have not occurred to anyone—at least, not long enough to write it down with mathematical precision—until Cardano began writing *The Book of Games of Chance.* Cardano developed subtle mathematical equations for analyzing dice games: for instance, he discovered the additive formula for calculating the odds of one of two potential events occurring in a single roll. (If you want to know how likely it is that you will roll either a three or an even number, you take the ⅙ likelihood of the three being rolled, and the ½ odds of the even number, and add them together. You will roll a three or an even number two out of three times on average with a six sided die.) He also demonstrated the multiplicative nature of probability when predicting the results of a sequence of dice rolls: the chance of rolling three sixes in a row is one in 216: ⅙ x ⅙ x ⅙.

Written in 1564, Cardano's book wasn't published for another century. By the time his ideas got into wider circulation, an even more important breakthrough had emerged out of a famous correspondence between Blaise Pascal and Pierre de Fermat in 1654. This, too, was prompted by a compulsive gambler, the French aristocrat Antoine Gombaud, who had written Pascal for advice about the most equitable way to predict the outcome of a dice game that had been interrupted. Their exchange put probability theory on a solid footing and created the platform for the modern science of statistics. Within a few years, Edward Halley (of comet legend) was using these new tools to calculate mortality rates for the average Englishman, and the Dutch scientist Christiaan Huygens and his brother Lodewijk had set about to answer "the question . . . to what age a newly conceived child will naturally live." Lodewijk even went so far as to calculate that his brother, then aged forty, was likely to live for another sixteen years. (He beat the odds and lived a decade beyond that, as it turns out.) It was the first time anyone had begun talking, mathematically at least, about what we now call life expectancy.

Probability theory served as a kind of conceptual fossil fuel for the modern world. It gave rise to the modern insurance industry, which for the first time could calculate with some predictive power the claims it could expect when insuring individuals or industries. Capital markets—for good and for bad—rely extensively on elaborate statistical models that predict future risk. "The pundits and pollsters who today tell us who is likely to win the next election make direct use of mathematical techniques developed by Pascal and Fermat," the mathematician Keith Devlin writes. "In modern medicine, future-predictive statistical methods are used all the time to compare the benefits of various drugs and treatments with their risks." The astonishing safety record of modern aviation is in part indebted to the dice games Pascal and Fermat analyzed; today's aircraft are statistical assemblages, with each part's failure rate modeled to multiple decimal places.

When we think about the legacy of Cardano and Pascal and Fermat, one question comes immediately to mind: What took us so long? Before Cardano, gamblers apparently had noticed that some results seemed more likely than others, but no one seems to have had the inclination or ability to explain *why* exactly. This is one of those blind spots that seems baffling to the modern mind. The Greeks developed Euclidean geometry and the Pythagorean theorem; the Romans completed brilliant engineering projects that stand to this day. Many of them played games of chance avidly; many of them had a significant financial stake in figuring out the hidden logic behind those games. Yet none of the ancients were able to make the leap from chance to probability. What held them back?

The answer to this riddle appears to lie with the physical object of the die itself. The *astragali* favored by the Egyptians and Greeks—along with other mechanisms for generating random results—were not the products of uniform manufacturing techniques. Each individual die would have its own idiosyncrasies. "[Each] had two rounded sides and only four playable surfaces, no two of which were identical," Devlin writes. "Although the Greeks did seem to believe that certain throws were more likely than others, these beliefs—superstitions—were not based on observation, and some were at variance with the actual likelihoods we would calculate today." Seeing the patterns behind the game of chance required random generators that were *predictable* in their randomness. The Greeks were playing with dice whereby some sets might favor the four over the one; others might be more inclined to land on a two. The unpredictable nature of the physical object made it harder to perceive the underlying patterns of probability.

All that had begun to change by the thirteenth century, when guilds of dice makers began to appear across Europe. Two sets of statutes from a dice-making guild in Toulouse, dating from 1290 and 1298, show how important the manufacture of uniform objects had become to the trade. The very first regulation forbids the production of "loaded, marked, or clipped dice" to prevent their association with swindlers and cheats. But the statutes also place a heavy emphasis on uniform design specifications: all dice had to have the exact same dimensions, with numbers positioned in the same configuration on the six sides of the cube. By the time Cardano picked up the game, dice had become standardized in their design. That regularity may have foiled the swindlers in the short term, but it had a much more profound effect that had never occurred to dice-making guilds: it made the *patterns* of the dice games visible, which enabled Cardano, Pascal, and Fermat to begin to think systematically about probability. Ironically, making the object of the die itself more uniform ultimately enabled people like Huygens and Halley to analyze the decidedly nonuniform experience of human mortality using the new tools of probability theory. No longer mere playthings, the dice had become, against all odds, tools for thinking.

The logic of games is ethereal. We have no idea how most ancient games were played, either because written rule books did not survive to modern times, or because the rules themselves evolved and then died out before the game's players adopted the technology of writing. But we know about these games because they were physically embodied in matter: in game pieces, sports equipment, even in rooms or arenas designed to accommodate rules that have long since been lost to history. We can see the rules embedded indirectly in the shapes of these artifacts, like the trace a mollusk leaves behind in a shell. Racquetball-like courts, constructed almost four thousand years ago by the Olmec civilization, which predates the Mayans and Aztecs, have been uncovered in modern-day Mexico. At roughly the same time, on the other side of the world, an Egyptian sculptor captured King Tuthmose III playing a ball game that resembled modern cricket. These artifacts—like the standardized dice of Toulouse—make it clear that games have not just tested our intellectual or athletic gifts: they've also tested our skills as toolmakers.

Invariably, the tools end up transforming the toolmakers, just as the regular shapes of those medieval dice helped their human creators think about probability in a new way. Consider one of humanity's original technologies: the ball. Some Australian Aborigines played with "a stuffed ball made from grass and beeswax, opossum pelt, or, in some cases, the scrotum of a kangaroo," writes historian John Fox. The Copper Inuits of the Canadian Arctic play a football-like game with a ball made from the hide of a seal. The Olmecs even buried their balls along with other religious talismans, the pre-Columbian equivalent of being buried in your Green Bay Packers jersey. Balls are the jellyfish of gaming evolution—at once ancient and yet still ubiquitous in the modern world. The Olmecs, Aztecs, and Mayans all failed to invent the wheel, but the ball was central to the culture of all three societies.

Ruins of a Mayan ball court, Honduras

During Columbus's second voyage to the Americas, the explorer and his crew observed the local tribes of Hispaniola (now Haiti) playing a ball game that may have been derived from the Olmec game. The spectacle of a sporting event would have been nothing new to European eyes at the end of the fifteenth century. But there was something captivating and mysterious about this game. The ball seemed to defy physics. Describing the ball's behavior several decades later, a Dominican friar wrote, "Jumping and bouncing are its qualities, upward and downward, to and fro. It can exhaust the pursuer running after it before he can catch up with it." Like the balls already omnipresent in Europe, the Hispaniola balls could be thrown with ease and would roll great distances. But these balls had an additional property. They could bounce.

Columbus and his crew didn't realize it at the time, but they were the first Europeans to experience the distinctive properties of

Christopher Columbus observes a ball game in Hispaniola

the organic compound isoprene, the key ingredient of what we now call rubber. The balls had been formed out of naturally occurring latex, a white sticky liquid produced by many species of plants, including one known as *Castilla elastica.* Around 1500 BC, the Mesoamerican natives hit upon a way to mold and stabilize the liquid into the shape of a sphere, which then possessed a marvelous elasticity that made it ideal for games. (They also used the material for sandals, armor, and raingear.) Rubber ball games became a staple of the Mesoamerican civilization for thousands of years, played by Mayans and Aztecs as well as the indigenous populations of Caribbean islands like Hispaniola. The games were both sporting events and religious rituals, officiated by priests and featuring idols that represented the gods of gaming. Some scholars believe the games were occasionally accompanied by ritual sacrifices as well.

Viewing the Haitians playing their games, Columbus and his crew allegedly found the elasticity of the balls so mesmerizing that they brought one back to Seville, though evidence of this is somewhat shaky. But a 1528 drawing by Christoph Weiditz marks the earliest verifiable appearance of these rubber balls in a European context. From his first voyage to the New World, Hernan Cortés had brought back two Aztec ballplayers, skilled athletes in a sport known as *ullamaliztli*, a hybrid word that unites the Aztec terms for ball play and rubber. The sport involved bouncing the balls off the hips and the buttocks, although variants permitted the use of hands and sticks. The athletes performed in the court of Charles V, where Weiditz sketched them at play, their mostly naked bodies protected only by loincloths and leather bands covering their posteriors. The Aztec ballplayers had been essentially kidnapped by Cortés, so we should be wary of making light of their appearance in the Spanish court, but their performance before Charles did mark the beginning of a practice that would become commonplace in the modern age: athletes imported from one part of the world to another thanks to their agility at playing with rubber balls.

The most important legacy of those *ullamaliztli* players would not belong to the world of sport, however. The real innovation lay in the rubber itself. Europeans first took notice of this puzzling new material thanks to the astonishing kinetics of the Mesoamerican balls. Several decades after the first balls made their way to Europe, the Spanish royal historian Pedro Mártir d'Angleria wrote, "I don't understand how when the balls hit the ground they are sent into the air with such incredible bounce?" The great seventeenth-century historian Antonio de Herrera y Tordesillas included a long account of the Mesoamerican "gum balls," which were "made of the Gum of a Tree that grows in hot Countries, which having Holes made in it distills great white Drops, that soon harden and being work'd and moulded together turn as black as pitch."

Scientists began experimenting in earnest with the material in the eighteenth century, including the British chemist Joseph Priestley, who allegedly noticed that the material was also uniquely suited for erasing (or "rubbing out") pencil marks, thus coining the name "rubber" itself. (Priestley had nothing to do with condoms, however.) Today, of course, the rubber industry is massive; we walk with shoes made with rubber soles, chew gum made from rubber compounds, drive cars and fly planes supported by rubber tires. The history of rubber's ascent is marred by exploitation, both of human and natural resources. "The demand for rubber was to see men carve out huge plantations from tropical forests around the world," John Tully writes in his social history of rubber, *The Devil's Milk*. "Roads and railways would be built, along with docks, power stations, and reticulated water supplies. In the process, whole populations would be transplanted to provide the labor power, and the ethnic composition of whole countries would be changed forever, as in Singapore and Malaysia. The demand was to lead to barbarism, as in the Belgian Congo and on the remote banks of the Putumayo River in Peru." Vast fortunes would be made out of isoprene's unique chemistry. Some of the most famous names in the history of Big Industry began their careers manufacturing and selling rubber products: Firestone, Pirelli, Michelin. Columbus had returned to Seville disappointed by his failure to bring back gold. He had no idea that he had stumbled across a material that would prove to be just as valuable—and far more versatile—in its eventual applications.

Today, of course, the canonical story of rubber innovation is that of the struggling nineteenth-century entrepreneur Charles Goodyear, who hit upon a technique (called vulcanization) that made rubber durable enough for industrial use, and made Goodyear himself a titan of industry. But the Mesoamericans had developed a vulcanization method thousands of years before Goodyear began his experiments. The prominence of the Goodyear narrative is partly due to a long-standing bias towards Euro-American characters in the history of innovation, but I suspect it also derives from another, more subtle bias: the assumption that important innovations come out of "serious" research like Goodyear's, fueled by entrepreneurial energy. But long before Goodyear's investigation, the Mesoamericans took the opposite path, driven not by industrial ambition but rather by delight and wonder. The rubber balls of the Olmecs make it clear that games do not just help concoct new metaphors or ways of imagining society. They can also drive advances in materials science. Sometimes the world is changed by heroic figures deliberately setting out to reinvent an industry and making a fortune in the process. But sometimes the world is changed just by following a bouncing ball.

In the fall of 1961, three MIT grad students were rooming together at what they would later recall as a "barely habitable tenement" on Hingham Street in Cambridge. (As a joke, they dubbed their living quarters the "Hingham Institute.") They were avid science-fiction readers, budding mathematicians, and steam-train aficionados. They would entertain themselves by watching low-budget Japanese monster movies at seedy Boston theaters. But most of their time they spent thinking about computers and their potential. Today we would call them hackers, though at the time the category didn't exist.

At that point in history, the sexiest piece of digital technology was a new machine from Digital Equipment Corporation: the PDP-1, considered at the time to be one of the first minicomputers ever released—although, to the modern eye, *mini* seems almost laughable, since it was the size of an armoire. What made the PDP-1 especially alluring to the fellows of the Hingham Institute was an accessory called the Type 30 Precision CRT: a circular black-and-white display. To the average American in 1961, who would have been just hearing the first hype about the coming revolution of color TV, the Type 30 would have looked like a step backward, but the fellows of the Hingham Institute knew the grainy images of the Type 30 augured a more radical transformation than color. The pixels on the Type 30 might have been woefully monochromatic, but you could control them with *software*. That made all the difference. For most of the short history of computing, human-machine interactions had traveled through the middleman of paper, through punch cards and printouts. You typed something in and waited for the machine to type something back. But with a CRT beaming electrons at a screen, that interaction became something fundamentally different. It was *live*.

This was more than three decades before Michael Dell pioneered the just-in-time manufacturing model for digital computers; ordering a PDP-1 was more like commissioning a yacht than it was one-click ordering a new gadget online. And so, when MIT announced that it was going to be installing a PDP-1 and Type 30 in the engineering lab, the fellows of the Hingham Institute had more than a few months to contemplate what they would do with such a powerhouse. As early adopters in an emerging academic discipline, they were cursed with the added burden of having to justify their field while simultaneously making progress in it. And so they began to think about applications that would both advance the art of computer science and dazzle the uninitiated. They came up with three principles:

1. It should *demonstrate* as many of the computer's resources as possible, and tax those resources to the limit.
2. Within a consistent framework, it should be interesting, which means every run should be different.
3. It should involve the onlooker in a pleasurable and active way—in short, it should be a game.

Inspired by the lowbrow sci-fi novels of Edward E. Smith that featured a steady stream of galactic chase sequences, the Hingham fellows decided that they would christen the PDP-1 by designing a game of cosmic conflict. They called it Spacewar!

The rules of Spacewar! were elemental: two players, each controlling a spaceship, dart around the screen, firing torpedoes at each other. Even in its infancy, you can see the family resemblance to the 1970s arcade classic Asteroids, which was heavily inspired by Spacewar! The graphics, of course, were pathetic by modern standards: it looked more like a battle between a ghostly pair of semicolons than a scene from *Star Wars*. But something about the experience of controlling these surrogates on the screen was hypnotic, and word of Spacewar! began to flow out of MIT and across the small but growing subculture of computer scientists. As the digital pioneer Alan Kay put it, "The game of Spacewar! blossoms spontaneously wherever there is a graphics display connected to a computer."

Just as chess evolved new rules as it migrated across the continents a thousand years before, the world of Spacewar! gathered new features as it traveled from computer lab to computer lab. Gravity was introduced to the game; specially designed control mechanisms (anticipating modern joysticks) were produced; a hyperspace option whereby your craft became momentarily invisible became customary; new graphics routines—many of them focused on explosions—made the game more lifelike. An MIT programmer named Peter Samson wrote a program—memorably dubbed "Expensive Planetarium"—that filled the Spacewar! screen with an accurate representation of the night sky. "Using data from the *American Ephemeris and Nautical Almanac*," writes J. Martin Graetz, one of the original Hingham Fellows, "Samson encoded the entire night sky (down to just above fifth magnitude) between 22½°N and 22½°S, thus including most of the familiar constellations . . . By firing each display point the appropriate number of times, Samson was able to produce a display that showed the stars at something close to their actual relative brightness." Integrated merely as a background flourish for Spacewar!, Expensive Planetarium marked one of the first times that a computer graphics program had modeled a real-world environment; when we look at family photos on our computer screens, or follow directions layered over Google Maps satellite images on our phones, we are interacting with descendants of Samson's flickering planetarium.

Dan Edwards and Peter Samson playing Spacewar! on the PDP-1 30 display

It goes without saying that Spacewar! began a lineage that would ultimately evolve into the modern video game industry, which generates more than $100 billion in sales annually. This is impressive enough, but perhaps not so surprising. Video games had to start somewhere, after all. But the legacy of Spacewar! extends far beyond PlayStation and Donkey Kong and SimCity. Just as Expensive Planetarium inaugurated a profound shift in the relationship between on-screen data and the real world, Spacewar! itself planted the seeds for a number of crucial developments in computing. The idea that a software application might be codeveloped by dozens of different programmers at different institutions spread around the world—each contributing new features or bug fixes and optimized graphics routines—was unheard-of when the Hingham Institute first began dreaming of space torpedoes. But that development process would ultimately create essential platforms like the Internet or the Web or Linux. Spacewar! was one of the first programs that successfully engaged the user in a real-time, visual interaction with the computer, with on-screen icons moving in sync with our physical gestures. Everyone who clicks on icons with a mouse or trackpad today is working within a paradigm that Spacewar! first defined more than half a century ago.

But perhaps the most fundamental revolution that Spacewar! set in motion was this: the game made it clear that these massive, unwieldy, bureaucratic machines could be hijacked by the pursuit of fun. Before Spacewar!, even the most impassioned computer evangelist saw them belonging to the world of Serious History: tabulating census returns, calculating rocket trajectories. For all their brilliance, none of the early visionaries of computing—from Turing to von Neumann to Vannevar Bush—imagined that a million-dollar machine might also be useful for blowing up an opponent's spaceship with imaginary torpedoes for sheer amusement. You might teach a computer to play chess in order to determine how intelligent the machine had become, but programming a computer to play games just for the sake of playing games would have seemed like a colossal waste of resources, like hiring a symphony orchestra to play "Chopsticks." But the Spacewar! developers saw a different future, one where computers had a more personal touch. Or, put another way, developing Spacewar! helped them see that future more clearly.

In 1972, during a hiatus between publishing issues of *The Whole Earth Catalog*, Stewart Brand visited the Artificial Intelligence Lab at Stanford to witness "the First Intergalactic Spacewar! Olympics." He wrote up his experiences for *Rolling Stone* in an article called "Spacewar: Fanatic Life and Symbolic Death Among the Computer Bums." As one of the first essays to document the hacker ethos and its connection to the counterculture, it is now considered one of the seminal documents of technology writing. In Spacewar!, Brand saw "a flawless crystal ball of things to come in computer science and computer use." The bearded computer "bums" tinkering with the rules to their sci-fi fantasy world were not just freaks and geeks; they offered a glimpse of what mainstream society would be doing in two decades. Brand famously began the essay with the proclamation "Ready or not, computers are coming to the people." In the sheer delight and playfulness of Spacewar!, Brand recognized that these seemingly austere machines would inevitably be domesticated and brought into the sphere of everyday life. Right after Brand's manifesto was published, another young hippie from the Bay Area started working for the pioneering game company Atari (which had been founded to bring a commercial version of Spacewar! to the market) and then shortly thereafter created a company devoted exclusively to manufacturing personal computers. His name, of course, was Steve Jobs.

In his *Rolling Stone* piece, Brand alluded to an important distinction between "low-rent" and "high-rent" forms of research. High-rent is official business: fancy R&D labs funded by corporate interests or government grants, reviewed by supervisors, with promising ideas funneled onto production lines. The world of games, however, is low-rent. New ideas bubble up from below, at the margins, after hours; people experiment for the love of it, and they share their experiments because they have none of the usual corporate restrictions that protect intellectual property. By sharing, they allow those new ideas to be improved by others along the chain. Before long, a game of juvenile intergalactic conflict becomes the seedling of a digital revolution, or a game of dice lays the groundwork for probability theory, or a board game maps out a new model of social organization. The ideas evolve in a low-rent world, but they often end up transforming the classier neighborhoods along the way.

One day in the late summer of 1961, just as the Hingham Institute was forming across the country in Cambridge, a woman stood at a roulette table in Las Vegas, placing bets in the chaos of the casino floor. At her side, a neatly dressed thirty-year-old man pushed his chips onto the betting area with a strange inconsistency. The man would never bet until a few seconds after the croupier had released the ball; many times he failed to bet at all, waiting patiently tableside until the next spin. But the pile of chips slowly growing in front of him suggested that his eccentric strategy was working. At one point, the woman turned to glance at the stranger beside her. A look of alarm passed across her face as she noticed a wiry appendage protruding from his ear, like the antennae of an "alien insect," as the man would later describe it. Seconds later, he was gone.

The mysterious stranger at the roulette table was not, contrary to appearances, a criminal or a mafioso; he was not even, technically speaking, cheating at the game—although years later his secret technique would be banned by the casinos. He was, instead, a computer scientist from MIT named Edward Thorp, who had come to Vegas not to break the bank but rather to test a brand-new device: the very first wearable computer ever designed. Thorp had an accomplice at the roulette table, standing unobserved at the other end, pretending not to know his partner. He would have been unrecognizable to the average casino patron, but he was in fact one of the most important minds of the postwar era: Claude Shannon, the father of information theory and one of the key participants in the invention of digital computers.

Thorp had begun thinking about beating the odds at roulette as a graduate student in physics at UCLA in 1955. Unlike card games like blackjack or poker where strategy could make a profound difference in outcomes, roulette was supposed to be a game of pure chance;

the ball was equally likely to end up on any number on the wheel. Roulette rotors with small biases could make certain outcomes slightly more likely than others. (Humphrey Bogart's casino in *Casablanca* featured a rigged roulette wheel that favored certain numbers.) But most modern casinos went to great lengths to ensure their roulette wheels generated perfectly random results. Using probability equations based on Cardano's dice analysis, casinos were then able to establish odds that gave the house a narrow, but predictable, advantage over the players. Thanks to his physics background, Thorp began to think of the roulette wheel's unbiased perfection as an opportunity for the gambler. With a "mechanically well made and well maintained" roulette wheel, Thorp later recalled, "the orbiting roulette ball suddenly seemed like a planet in its stately, precise and predictable path." With astral bodies, of course, knowing the initial position, direction, and velocity of the object would enable you to predict its location at a specified later date. Could you do the same thing with a roulette ball?

Thorp spent some time tinkering with a cheap, half-sized roulette wheel he had acquired, analyzing ball movements with a stopwatch accurate to hundredths of a second, even making slow-motion films of the action. But the wheel proved to be too defective to make accurate predictions, and Thorp got sidetracked writing software programs to calculate winning strategies at blackjack. (One gets the sense that Cardano and Thorp would have hit it off nicely.) In 1960, having moved on to MIT, Thorp decided to try publishing his blackjack analysis in the *Proceedings of the National Academy of Sciences*, and sought out the advice of Shannon, the only mathematician at MIT who was also a member of the academy. Impressed by Thorp's blackjack system, Shannon inquired whether Thorp was working on anything else "in the gambling area." Dormant for five years, Thorp's roulette investigation was suddenly reawakened as the two men began a furious year of activity, seeking a predictable pattern in the apparent randomness of the roulette wheel.

Claude Shannon with an electronic mouse

In his old, rambling wooden house outside of Cambridge, Shannon had created a basement exploratorium that would have astounded Merlin and Babbage. Thorp later described it as a "gadgeteer's paradise":

> *It had perhaps a hundred thousand dollars (about six hundred thousand 1998 dollars) worth of electronic, electrical and mechan-*

> *ical items. There were hundreds of mechanical and electrical categories, such as motors, transistors, switches, pulleys, gears, condensers, transformers, and on and on. As a boy science was my playground and I spent much of my time building and experimenting in electronics, physics and chemistry, and now I had met the ultimate gadgeteer.*

Before long, Thorp and Shannon had spent thousands of dollars acquiring a regulation wheel from Reno, along with strobe lights and a specialized clock. They commandeered an old billiards table in Shannon's house and began analyzing the orbit of the ball on a spinning roulette wheel. Just as Thorp had intuitively surmised five years before, the ultimate destination of the ball could be predicted with some accuracy if you had a way of measuring its initial velocity. The two men decided the best way to assess that velocity would be to measure the length of time it took the ball to complete one rotation around the wheel. A long rotation would suggest a slower velocity, while a short rotation suggested that the ball was traveling at a higher speed. That initial velocity assessment gave them an indication of when the ball would eventually fall off the sloped track and onto the rotor, eventually settling into the numbered frets. The higher the velocity, the longer it would take to drop onto the rotor. Getting a read on the initial velocity couldn't tell you exactly what number the ball would eventually land on, but it did suggest a sector of the wheel where the ball was slightly more likely to come to a stop. Because the house's advantage in roulette is set at such a narrow margin, even a slight edge in predicting the ball's outcome could tip the advantage in the player's favor.

The problem was you couldn't exactly stroll onto the casino floor at the Palms with a strobe light and stopwatches. To use Shannon and Thorp's technique, you needed some way to measure the velocity of the ball in real time without anyone else at the roulette table noticing. And you needed to calculate how tiny differences in velocity would translate into the ball's ultimate resting place. You needed some kind of sensor to track the ball's physical movements, and you needed a computer to do the math. The prospects of this imagined system would have been laughable to anyone familiar with the technological state of the art in 1961. The smallest video camera in the world was the size of a suitcase, and most computers were larger than a refrigerator. Thorp and Shannon wouldn't have been vulnerable to getting caught cheating in the Vegas casinos with that gear. They would have never made it in the front door.

But Shannon in particular wasn't just familiar with the technological state of the art. He was one of its masters. His approach to innovation anticipated the rec-room informality that would come to define the Google or Facebook campuses decades later; work and play were inextricably linked. In Thorp's description:

> *As we worked and during breaks, Shannon was an endless source of playful ingenuity and entertainment. He taught me to juggle three balls (in the '70's he proved "Shannon's juggling theorem") and he rode a unicycle on a "tightrope," which was a steel cable about 40 feet long strung between two tree stumps. He later reached his goal, which was to juggle the balls while riding the unicycle on the tightrope. Gadgets and "toys" were everywhere. He had a mechanical coin tosser which could be set to flip the coin through a set number of revolutions, producing a head or tail according to the setting. As a joke, he built a mechanical finger in the kitchen which was connected to the basement lab. A pull on the cable curled the finger in a summons.*

Eventually, between the juggling sessions and the unicycle tightrope rides, Shannon and Thorp designed a computer with twelve transistors that could be concealed in a box the size of a deck of cards. Instead of using cameras to track the ball's motion, they designed special input mechanisms in their shoes; using their toes to activate microswitches that marked the beginning and end of the ball's initial rotation around the wheel. Based on that assessment of the ball's velocity, the microcomputer calculated the most likely landing spot, conveying the information with one of eight musical tones, corresponding to eight sectors of the roulette wheel. The mix of technology Thorp and Shannon assembled to crack the roulette game had not only never been seen before, it had barely been imagined. Even the most visionary computer scientists or science-fiction authors imagined computers as bulky, stationary objects—closer to a piece of furniture than a piece of jewelry. But fifty years later, Thorp and Shannon's roulette-wheel hack would become the most ubiquitous form of computing on the planet: a small digital device in your pocket, attached to headphones, with sensors recording your body's movements. It might have looked like two men goofing off and trying to beat the house at roulette, but it was also something much more profound. An entire family tree of devices—iPods, Android phones, Apple Watches, Fitbits—descend directly from that roulette hack.

In the end, Shannon and Thorp ventured to Vegas with their wives and gave their system a test run with real money. Shannon usually stood by the croupier and timed the ball with his toes; Thorp placed bets on the other end of the table while the wives stood lookout for any undue attention from the casino personnel. (Everyone but Thorp was nervous about being discovered by the mob element that controlled much of Vegas in those days.) They only bet ten-cent chips, but reliably beat the house. Their novel microcomputer performed brilliantly; ironically, the components of the system that proved to be the most failure-prone were the earphones.

After the Vegas trip, Thorp left MIT and published a few popular books on blackjack technique. Shannon and Thorp revealed the details of their secret invention in 1966, inspiring a few imitators over their years with more pecuniary objectives. Microdevices that monitored blackjack and roulette wheels were finally banned in the state of Nevada in 1985. But by that time, Thorp had little need for roulette winnings, having found a bigger casino to exploit. Using his mathematical skills and his experience discovering small opportunities in probability, he founded a company in 1974 called Princeton/Newport Partners. It was one of the first of a new genus of finance firms that would eventually become a wellspring of profit *and* controversy: the hedge fund.

In the mid-2000s, the head of research at IBM, Paul Horn, began thinking about the next chapter in IBM's storied tradition of "Grand Challenges"—high-profile projects that showcase advances in computation, often with a clearly defined milestone of achievement designed to attract the attention of the media. Deep Blue, the computer that ultimately defeated Gary Kasparov at chess, had been a Grand Challenge a decade before, exceeding Alan Turing's hunch that chess-playing computers could be made to play a tolerable game. Horn was interested in Turing's more celebrated challenge: the Turing Test, which he first formulated in a 1950 essay on "Computing Machinery and Intelligence." In Turing's words, "A computer would deserve to be called intelligent if it could deceive a human into believing that it was human."

The deception of the Turing Test had nothing to do with physical appearances; the classic Turing Test scenario involves a human sitting at a keyboard, engaged in a text-based conversation with an unknown entity who may or may not be a machine. Passing for a human required both an extensive knowledge about the world and a natural grasp of the idiosyncrasies of human language. Deep Blue could beat the most talented chess player on the planet, but you couldn't have a conversation with it about the weather. Horn and his team were looking for a comparable milestone that would spur research into the kind of fluid, language-based intelligence that the Turing Test was designed to measure. One night, Horn and his colleagues were dining out at a steak house near IBM's headquarters and noticed that all the restaurant patrons had suddenly gathered around the televisions at the bar. The crowd had assembled to watch Ken Jennings continue his legendary winning streak at the game show *Jeopardy!*, a streak that in the end lasted seventy-four episodes. Seeing that crowd forming planted the seed of an idea in Horn's mind: Could IBM build a computer smart enough to beat Jennings at *Jeopardy!*?

The system they eventually built came to be called Watson, named after IBM's founder Thomas J. Watson. To capture as much information about the world as possible, Watson ingested the entirety of Wikipedia, along with more than a hundred million pages of additional data. There is something lovely about the idea of the world's most advanced thinking machine learning about the world by browsing a crowdsourced encyclopedia. (Even H. G. Wells's visionary prediction of a "global brain" didn't anticipate that twist.) From the data sources, and from software prompting by the IBM researchers, Watson developed a nuanced understanding of linguistic structures that let it parse and engage in less rigid conversations with humans. By the end of the process, Watson could understand and answer questions about just about anything in the world, particularly answers that relied on factual statements. And of course, given the Grand Challenge that had inspired its creation, Watson knew how to present those answers in the form of a question.

In February of 2011, Watson competed in a special two-round *Jeopardy!* competition against Jennings and another *Jeopardy!* champion named Brad Rutter. Viewers at home were treated to a glimpse inside Watson's "mind" as it searched its databases for the correct answer to each question, via a small graphic that showed the computer's top three guesses and overall "confidence level" for each answer. Somewhat disingenuously, the IBM promotional video that aired at the beginning of the episode suggested that Watson was like the two other human players in that it "did not have access to the Internet," which is a bit like saying you don't have "access" to the cake you just consumed.

Watson made some elemental mistakes during the match—at one point answering "What is Toronto?" in a category called "U.S. Cities." But in the end, the computer trounced its human opponents, almost doubling the combined score of Jennings and Rutter. Watson's performance was particularly impressive given the rhetorical subtlety of *Jeopardy!*'s questions. In the second match, Watson correctly answered "What is an event horizon?" with 97 percent confidence for a question that read, "Tickets aren't needed for this 'event,' a black hole's boundary from which matter can't escape." To make sense of that question, Watson had to effectively disregard the entire first half of the statement, and recognize that the allusion to "tickets" is simply a play on another meaning of *event*. The final question that sealed Watson's victory was a literary one: "An account of the principalities of Wallachia and Moldavia inspired this author's most famous novel." Watson got it right: "Who is Bram Stoker?" Jennings had the answer as well, but recognizing that he had just lost to the machine, he scribbled a postscript beneath his answer: "I for one welcome our new computer overlords." The computer overlords may have a good chuckle over that one someday.

In terms of sheer computational power, many other supercomputers around the world outperform Watson. But because of Watson's facility with natural language, and its ability to make subtle inferences out of unstructured data, Watson comes as close to the human experience of intelligence as any machine on the planet. As Jennings later described it in an essay:

> *The computer's techniques for unraveling* Jeopardy! *clues sounded just like mine. That machine zeroes in on key words in a clue, then combs its memory (in Watson's case, a 15-terabyte data bank of human knowledge) for clusters of associations with those words. It rigorously checks the top hits against all the contextual informa-*

tion it can muster: the category name; the kind of answer being sought; the time, place, and gender hinted at in the clue; and so on. And when it feels "sure" enough, it decides to buzz. This is all an instant, intuitive process for a human Jeopardy! *player, but I felt convinced that under the hood my brain was doing more or less the same thing.*

IBM, of course, has plans for Watson that extend far beyond *Jeopardy!*. Already Watson is being employed to recommend cancer treatment plans by analyzing massive repositories of research papers and medical data, and answer technical support questions about complex software issues. But still, Watson's roots are worth remembering: arguably the most advanced form of artificial intelligence on the planet received its education by training for a game show.

You might be inclined to dismiss Watson's game-playing roots as a simple matter of publicity: beating Ken Jennings on television certainly attracted more attention for IBM than, say, Watson attending classes at Oxford would have. But when you look at Watson in the context of the history of computer science, the *Jeopardy!* element becomes much more than just a public relations stunt. Consider how many watershed moments in the history of computation have involved games: Babbage tinkering with the idea of a chess-playing "analytic engine"; Turing's computer chess musings; Thorp and Shannon at the roulette table in Vegas; the interface innovations introduced by Spacewar!; Kasparov and Deep Blue. Even Watson's 97 percent confidence score in the "event horizon" answer relied on the mathematics of probability that Cardano had devised to analyze dice games five hundred years before. Of all the fields of human knowledge, only a handful—primarily mathematics, logic, and electrical engineering—have been more important to the invention of modern computers than game play. Computers have long had a reputation for being sober, mathematical machines—the Vulcans of the technology world—but the history of both music and games, from The

Instrument that Plays Itself to Spacewar!, make it clear that the modern digital computer has a long tradition of play and amusement in its family tree. That history may explain in part why computers have become so ubiquitous in modern life. The abacus and the calculator were good at math too, but they didn't have the knack for wonder and delight that computers possess.

Why were games so important to the history of computing? I suspect the answer is partially that games offer a clearly defined yardstick or score by which progress can be measured. Beating the house at roulette, or Ken Jennings at *Jeopardy!*, gives the scientists and inventors a definitive goal that helps clarify and direct research. A more important principle is also visible in the long alliance between games and computation. Playing rule-governed games is one of those rare properties of human behavior that seems to belong to us alone as a species. So many other forms of human interaction and cognition have analogs in other creatures: song, architecture, war, language, love, family. Rule-governed game play is both ancient and uniquely human. It is also one of the few activities—beyond the essentials like eating, sleeping, and talking—that three-year-olds and ninety-three-year-olds will all happily embrace. But unlike those essentials, there is no clear evolutionary purpose for game play. It does not provide carbohydrates to burn, or help us reproduce our genes, yet somehow our minds are drawn compulsively to the challenge and unpredictability of games. And so it seems fitting that when those minds finally got smart enough to think about building an *artificial* mind, one of the first challenges we set for ourselves was creating a machine that could play along.

6

Public Space

The Pleasuring Grounds

John Hughson was a down-on-his-luck cobbler when he decided, in the mid-1730s, to make a pilgrimage from Yonkers to Manhattan. He set up shop in the big city, then sporting a population of around twenty thousand people. Hughson opened a series of taverns, first by the banks of the East River, and then on the Hudson, near the site where the World Trade Center would rise, and fall, almost three centuries later. The tavern on the West Side soon developed a reputation as a popular watering hole—not just for the booze, but also for a regular "great feast" Hughson would host each Sunday. A crowd would start gathering after church services wound down, and Hughson would set out a table of mutton and goose, accompanied by rum, cider, and beer. "Someone might play a fiddle, and often there was dancing," one historian writes. "There was gambling, too. Dice were rolled for prizes of bowls of punch, or wagers were laid on cockfights." The great feasts were a time to "frolick and merry make," a regular at Hughson's later recalled. Some of that "merrymaking" involved prostitution as well; Hughson's establishment was renowned for one prostitute in particular, a woman named

Margaret Sorubiero, who seems to have gone by a string of aliases, but was most commonly known as "the Newfoundland Irish Beauty."

In all, Hughson's tavern was like many public houses or drinking establishments across colonial America or Europe at the time: a raucous, festive enclave with somewhat lax moral standards. But a visitor to Hughson's would notice one key difference within seconds of walking into the tavern: while Hughson himself was white, his clientele was largely made up of African-American slaves mingling with a few lower-class whites, some of them soldiers. (The historians Edward G. Burrows and Mike Wallace suggest that the crowd also included some "young gentlemen inclined to after-hours carousing and gaming.") While slavery wasn't officially abolished in New York until 1827, taverns catering to slaves were not unheard-of in the city. A joint called Cato's Road House, founded by a former South Carolina slave, stayed in business for fifty years during the first half of the eighteenth century, supplying brandy and cigars to slaves on the run from their southern masters. But the mix of white and black customers set Hughson's establishment apart. Perhaps most scandalously, "the Newfoundland Irish Beauty" had openly had a child with a slave known as Caesar. The casual intermingling of races was otherwise nonexistent in eighteenth-century Manhattan; despite the city's reputation for tolerance, the lines of racial segregation were clearly drawn. But at Hughson's, those lines began to blur, with ultimately catastrophic consequences.

Starting in March of 1741, a series of suspicious fires erupted in Manhattan, beginning with a blaze at the governor's house in Fort George, followed by fires at warehouses, private homes, and stables across the city. Word began to spread through the white population that the fires were being started by rebellious slaves; when an African-American slave named Cuffee was seen running from a blaze in early April, he was quickly arrested and thrown in jail. Vigilante crowds soon began hunting down other slaves and handing them over to the authorities. As the panic spread, a grand jury, supervised by a justice named Daniel Horsmanden, was convened to crack down on the white proprietors who were allegedly inciting revolt by supplying alcohol to African-Americans, which naturally brought the authorities to John Hughson.

As it turned out, Hughson was already in trouble with the law. Alongside the gambling and the prostitution, he apparently ran a profitable side business fencing stolen property out of his tavern. (The tavern, for all we know, may have actually been the side business, with the fencing the main moneymaker.) The local police had long suspected that Hughson was part of a larger crime organization, but hadn't been able to find convincing evidence of his wrongdoing, until a sixteen-year-old indentured servant of Hughson's named Mary Burton decided to inform on her boss after the authorities offered to free her from her obligations to him. Initially, Burton merely accused Hughson of conspiring with Caesar in a recent robbery of a general store on Broad Street. But after the fires broke out, Burton took her role of accommodating witness to a whole new level, becoming, as the historian Christine Sismondo puts it, "more than happy to give up the name of everybody involved in the robbery, the fencing, and, it would turn out, just about anyone whose name she'd ever even heard." Burton reported that she had overhead numerous conversations during which Hughson, Caesar, and Cuffee actively discussed inciting a slave rebellion by leaving the city in ashes. Hughson allegedly told Caesar that in the new regime, the two men would be "King" and "Governor" respectively. (Many historians today believe the two men were discussing leadership of a crime syndicate, not a post-rebellion slave colony.) Burton also testified that she had overhead slaves at Hughson's tavern exclaiming, "God damn all the white people."

Burton's testimony triggered one of the most brutal eruptions of state violence in the history of New York. Horsmanden's grand jury swiftly convicted Caesar of burglary, and on May 11, he and another slave were hanged in gallows near Freshwater Pond. The authorities left his body hanging from the scaffolding for weeks as a savage reminder of the fate that awaited other rebellious slaves. Within a matter of weeks, Cuffee and another alleged conspirator were sentenced to death by fire. They attempted to buy themselves a more lenient sentence by giving away the names of other supposed conspirators (many of whom were eventually hanged), but their last-minute confession did them no good. They were burned alive in front of an enraged mob.

The grand jury then turned to Hughson, who, along with his wife and the Newfoundland Irish Beauty, were quickly sentenced to hang as well. The trial made it clear that Hughson's true offense lay in the social space he had created in the tavern itself, with its scandalous mingling of black and white and its reversal of traditional race relations. During the proceedings, one justice asserted that Hughson had broken the law simply by selling "a penny dram of a penny worth of rum to a slave, without the direct concept or direction of his master." During the sentencing, another judge made the case against Hughson's integrationist tendencies in the strongest language imaginable:

> *For people who have been brought up and always lived in a Christian country, and also called themselves Christians, to be guilty not only of making Negro slaves their equals, but even their superiors, by waiting upon, keeping with, and entertaining them with meat, drink and lodging, and what much more amazing, to plot, conspire, consult, abet and encourage these black seed of Cain, to burn this city, and to kill and destroy us all. Good God! When I reflect on the disorders, confusion, desolation and havoc, which the effect of your most wicked, most detestable and diabolical councils might have produced, (had not the hand of our great and good God interposed) it shocks me!*

In the end, after executing the two women alongside Hughson, the authorities strung up Hughson's body next to the decomposing corpse of Caesar. A bizarre rumor spread through the city that, in death, Caesar's skin had turned white, while Hughson's had turned black, a macabre end to a story that had, from the beginning, revolved around the scandalous interactions of white and black skin.

The turbulent events of that spring are generally referred to today as the "Slave Rebellion of 1741" or the "New York Conspiracy of 1741." But the truth is no one knows for certain whether there was a rebellion or a conspiracy at all. The most plausible scenario is that Hughson's accomplices may have been setting fires as a distraction while they stole goods from nearby stores and homes. But it's also possible that there was no connection between Hughson and the fires, and the whole thing was the creation of a sixteen-year-old girl's imagination, enticed into her accusations by the promise of freedom. But one fact in this entire murky business remains unchallenged: Hughson's tavern was guilty—in the eyes of the New York establishment—of expanding the boundaries of legitimate social interaction between the races. Whatever illegal activities were perpetrated in Hughson's tavern, the community that formed in that environment was, in a real sense, a century or two ahead of its time. Today, in most parts of the United States, we wouldn't think twice about a bar that attracted a multiracial clientele. But in 1741 it was apostasy. The most telling clues about the future of racial integration were not visible in the homes or churches or businesses of New York. If you wanted to see where race relations were headed, you had to go to a bar.

The tavern is an ancient institution, almost as old as civilization itself. We have no record of the cunning entrepreneur who first came up with the idea of carving out a space where drinks could be purchased, and camaraderie enjoyed, with a group of friends and strangers. Most likely the idea was independently developed by many people at different places and times. (Once you had beer, beer gardens inevitably

Roman* tabernae *ruin in Pompeii

beckoned.) When they first emerged, the taverns marked the origin point of a line that would lead not just to McSorley's, and Harry's Bar, and *Cheers*. The birth of the drinking house also marked the origins of a new kind of space: a structure designed explicitly for the casual pleasures of leisure time. The tavern was not a space of work, or worship; it was not a home. It existed somewhere else on the grid of social possibility, a place you went just for the fun of it. Our modern world is now teeming with these spaces: bars, cafés, spas, resorts, casinos, theme parks. Entire cities now brand themselves as entertainment experiences, escapes from the daily grind. The original taverns, wherever they took root, gave humans the first glimpse of that wonderland future, with all its glamour and its absurdity.

The birth of the tavern deserves a place in the history books as a symptom of rising standards of living. "Luxury goods" predate the first cities, in gems and other trinkets that passed through Neolithic society. But luxury *experiences* were something new. Creating an economic system where it was possible to drop down to the local bar and spend some spare cash on drinks for a few hours—this was no small achievement in itself. For the most part, we tolerate bars and pubs and taverns as guilty pleasures: yes, they attract (and encourage) illicit behavior, as Hughson's fencing ring and the Newfoundland Irish Beauty attest. But the sentimental, communal appeal of the local dive usually overrides that puritanical scolding. What we rarely acknowledge, however, is the transformative role taverns and bars have played in our *political* history. Your seventh-grade history book may have told you that democracy was born in the agora, where toga-wearing philosophers debated the issues of the day in the town square. But the truth is democracy, like countless scandalous offspring since, was just as likely to have been conceived in a bar.

The word *tavern* derives from the Roman *tabernae,* although the ancient Greeks had public drinking houses as well, where heavily watered-down wine was served to patrons. Archeologists cataloging the ruins of Pompeii discovered 118 distinct taverns in the town; the ash cloud from Vesuvius even preserved a chalkboard with a wine list inscribed on it: "For one [coin] you can drink wine; For two you can drink the best; For four you can drink Falernian." On one tavern wall, two-thousand-year-old graffiti complains about the quality of the booze: "Curses on you, Landlord, you sell water and drink un mixed wine yourself."

Since many *tabernae* also doubled as inns, Romans also deployed taverns as nodes on their vast imperial network, layover points that enabled men to travel thousands of miles with a reliable form of shelter—and a good drink—guaranteed almost every night. According to W. C. Firebaugh, one of the earliest historians of Greco-Roman tavern culture, "the stages of travel were so admirably calculated that the end of each day's journey found the traveller at a station where fresh horses and pack animals could be obtained, and where food and lodging were procurable." It wasn't enough to build a global network of roads, all leading to Rome; the Romans had to create a system whereby travelers could make it to the ends of the empire and back without relying on the kindness of local strangers, or sleeping in a field or forest. Roman planners built taverns every fifteen miles on the road, becoming a de facto unit of measuring distance. The tavern system was a key resource in managing the logistical nightmare of maintaining a global empire in an age before trains, planes, automobiles, or the Internet.

But however important the Roman *tabernae* may have been to the empire itself, their ultimate legacy lay in introducing the convention of public drinking houses to much of Europe. In England, those Roman taverns evolved into that most iconic of British institutions: the "publican" or "public house"—or as we now say, the pub. The census of 1577 reported the existence of more than sixteen thousand pubs in England, which suggested a ratio of one pub for every 187 residents. (Today, the ratio in the UK is closer to one pub per thousand citizens.) Not surprisingly, that level of pub density triggered much outcry from the more sober members of society. "The multiplying of taverns is evident cause of the disorder of the vulgar people," Secretary of State William Cecil warned, "who by haunting thereto waste their small substance which they weekly get by their hard labor and commit all evils that accompany drunkenness." But the pubs were not just inciting disorderly conduct; they also established a home for a new kind of conversation and community, what the historian Ian Gately calls "a fresh ethos":

> *[Pubs] were run by the people, for the people. They were places where men and women from different occupations and backgrounds might meet to drink and to enjoy each other's company, and where they might talk with candor about their rulers. Indeed,*

> *the common people enjoyed a freedom of speech and action in their drinking places that was denied to them elsewhere, and these institutions became the nucleus of a popular culture.*

The political impact of the drinking house would be felt most profoundly not in England but in her colonies. Ben Franklin once estimated that there was a tavern for every twenty-five men in Philadelphia—far exceeding the pub density of Britain. Just as Hughson's tavern played a defining role in the alleged slave rebellion of 1741, taverns throughout the American colonies were the seedbeds of the rebellion that would ultimately become the Revolutionary War. The legendary Boston Caucus that would eventually spawn the Boston Town Meeting and the Sons of Liberty first formed in a Boston tavern run by Elisha Cooke Jr. (The word *caucus* itself either derives from "Cook's House" or the Greek word for wine bowl.) The plans for the Boston Tea Party were hatched in another Boston tavern, the Green Dragon, which came to be known as "the headquarters of the American Revolution." A decisive independent streak ran through the tavern culture of the colonies; many taverns, like Hughson's, were used by smugglers trying to evade British taxes. (An underground tunnel connected the kitchen of a Philadelphia tavern to the city docks.) An analysis of voting records from the 1700s by historian David Conroy has shown that tavern keepers made up a disproportionate number of representatives elected to the Massachusetts House, causing John Adams to complain that the coarse and raucous taverns were "becoming in many places the nurseries of our legislators." Both Thomas Paine's *Common Sense* and the Declaration of Independence were read aloud in taverns throughout the colonies, stoking the fires of revolution.

It is difficult, in fact, to find a single momentous event in the decades leading up to the Revolutionary War that didn't unfold, in part, in the semipublic confines of a tavern. Erase those drinking establishments from history and it is entirely possible the radical front of the Sons of Liberty would have taken far longer to self-organize, and the revolution itself could have been postponed for decades. A revolutionary war with a fully industrialized Britain in the early 1800s might well have gone the other way.

It's worth pausing for a second to get the causality right here. American independence wasn't *caused* by the prevalence of tavern culture in the colonies. There were many forces at work, some of them likely stronger than the space of dissent that the tavern offered the early revolutionaries. But the existence of that space was nonetheless a determinate factor in the way the events unfolded. Change that one variable, and the buildup to the War of Independence has to, at the very least, unfold along a different path, since so much of the debate and communication relied on the semipublic exchange of the tavern: a space where seditious thoughts could be shared, but also kept secret. Like the Roman *tabernae* that preceded them, the American taverns were critical nodes on a network. The Romans used them to keep one empire together. The Americans used them to topple another.

The scene spilling out of the Black Cat Tavern in Los Angeles's Silver Lake neighborhood on the night of December 31, 1966, was precisely the sort of spirited chaos you might expect to find at an urban gay bar on New Year's Eve. A costume party had just ended at another bar down the street, and a trio of African-American men in drag—known as the Rhythm Queens—had launched into a upbeat rendition of "Auld Lang Syne." Gay bars like the Black Cat had become increasingly visible in American downtowns in the postwar years, though their roots extended back to nineteenth-century establishments, like Pfaff's Beer Cellar in Manhattan, the Bohemian enclave where Whitman sought out the "beautiful young men . . . with bright eyes." The bars were both pickup spots and a rare space where gay and lesbian people could be open about their sexual

The Green Dragon tavern, Boston

identity in an age when the closet was the only option. Like Pfaff's, they also served as networking hubs for the creative class: Tennessee Williams spent many hours at Café Lafitte in Exile in New Orleans; Allen Ginsberg held court in San Francisco's Black Cat.

But these bars were also spaces of challenge and confrontation. Mingling in the crowd that New Year's Eve night in Los Angeles were a dozen plainclothes officers from the LAPD Vice Squad, long the enemy of the city's gay population. As the balloons dropped, marking midnight, and the men in the Black Cat embraced and exchanged New Year's kisses, the undercover officers—assisted by a sudden influx of uniformed police bearing billy clubs—attacked and arrested sixteen men, beating several so badly that they had to be hospitalized. The undercover officers later testified that they had seen six men "kissing other men on the lips for up to ten seconds." All six were subsequently found guilty of "lewd conduct."

The raid at the Black Cat preceded the much more famous confrontation at the Stonewall Inn in the West Village of Manhattan two years later. But the events of that New Year's Eve night played a critical role in the formation of the gay rights movement. Six weeks after the raid, a street protest was staged on Sunset Boulevard, with hundreds of gay Angelenos chanting, "No more abuse of our rights and dignity." Inspired by the Black Cat affair, a group of activists that had recently formed an organization called Personal Rights in Defense and Education (PRIDE) decided to turn their biweekly mimeographed newsletter into a genuine newspaper. The result was the *Advocate*, the first national publication devoted exclusively to the gay and lesbian community—a publication that would eventually become an essential component of the many struggles and victories of the gay community of the ensuing decades: the post-Stonewall birth of gay liberation, the HIV/AIDS crisis, and the gay marriage movement.

You can think of the impact of the semipublic space of bars and clubs on the politics of gay liberation as a kind of cultural hummingbird effect. Many of the unlikely transformations we have surveyed here revolved around technological breakthroughs: someone invents a device specifically designed for a single purpose, but the introduction of that device into wider society triggers a series of changes that the inventor never dreamed of. A group of showmen and amateur scientists stumbles across a technique for fooling the eye into perceiving motion in a series of still images, and a century and a half later, that innovation has created a new class of celebrity, and a cottage industry devoted to nourishing the imaginary relationships between reality-show stars and their fans. But some of the most influential innovations in our history do not involve new machines or scientific principles. Bars and pubs and taverns might have used new technologies over the centuries—from corkscrews to kegs to refrigerators—but what made the drinking house so transformative had nothing to do with the contraptions it employed. The innovation, instead, was more social and physical: the idea of a space that was both open to the public but also closed off from the street, where one could comfortably alter one's mental state for a few hours. Over the long arc of history, this innovation would prove to be as important as the more traditional breakthroughs in the canon of good ideas: the sextant, say, or the pocket watch. The immediate purpose of the tavern was clear: it was an environment that made it easy for people to get drunk, and for the proprietors to make money by selling the drinks. But the hummingbird effect of the tavern turned out to be social and political: the tavern proved to be an environment that pushed the boundaries of social relationships, encouraging experimentation and nurturing dissent. The first person to hang out a shingle and serve drinks to paying customers—at some point back in the dawn of civilization—almost certainly had no idea that his or her innovation would ultimately support political and sexual revolutions that would reverberate around the world. A space originally intended for play and leisure became, improbably enough, a hotbed of dangerous new ideas.

These kinds of spaces played a defining role in one of the most influential works of sociology published in the twentieth century: Jürgen Habermas's *The Structural Transformation of the Public Sphere*. Originally conceived while Habermas was working as a graduate student under the legendary Frankfurt School Marxists Theodor Adorno and Max Horkheimer, the book was actually his doctoral dissertation, though Habermas broke off with his mentors due to their "paralyzing political skepticism." (With almost fifteen thousand citations tracked in Google Scholar, it stands as one of the most cited dissertations in the history of academia.) Habermas observed that, starting around the middle of the seventeenth century, the concept of "the public" (or *le public* in French, or *Publikum* in German) took on a new prominence in the languages of Western Europe. Before this point, people alluded to "the world" or "mankind" when talking about a general audience or crowd. But the idea of a public implied that there was a body of opinion and taste that possessed its own force and influence in society, potentially rivaling that of the monarchies and clergy. For the first time, people began talking explicitly about the court of "public opinion"; they began to seek "publicity" for their work or ideas, a word that originates with the French *publicité*. Habermas argued that the political and intellectual revolutions of the eighteenth century had been facilitated by the creation of this new public sphere, largely housed in semipublic gathering places like taverns and pubs. (A few decades after Habermas, the American sociologist Ray Oldenburg would develop a similar thesis in a book called *The Great Good Place*—coining the now-common expression "the third place" for these venues.) For Habermas, the public sphere had a profoundly egalitarian bias, creating "a kind of social intercourse that, far from presupposing equality of status, disregarded status altogether. [Participants] replaced the celebration of rank with a tact befitting equals." And, like the Black Cat and the Stonewall Inn two centuries later, it was a space that "presupposed the problematization of areas that until then had not been questioned." For Habermas, the public sphere was not simply architectural; it was also facilitated by new developments in media, particularly the rise of pamphleteering that was so central to Enlightenment discourse.

Taverns, salons, and drinking societies play a role in *The Structural Transformation of the Public Sphere*. (Had Habermas focused more on the American Revolution, they might have even had a starring role.) But for Habermas, the revolutionary ideas of the eighteenth century were ultimately dependent not on the beer and wine of tavern culture, but on another drug that had just arrived in the cities of Europe: coffee.

The story of how humans developed a taste for coffee—and, indirectly, an addiction to caffeine, now the most popular psychoactive compound on the planet—conventionally dates back to the Ethiopian city of Harar, where it is believed that the coffee plant, *Coffea arabica*, was first domesticated. But in a way, the story is much older than that. After all, the distant ancestors of *Coffea arabica* began producing caffeine in their berries long before humans noticed that the drug provided a pleasant jolt of mental stimulation. Modern genomics has now begun to explain why the plants evolved this potent chemical in the first place. In September of 2014, an international team of researchers, led by French and American scientists, announced that they had sequenced the genome of *Coffea canephora*, a plant that now accounts for about a third of the coffee consumed around the world. Analyzing sections of the genome that direct the production of caffeine, and comparing them to equivalent caffeine-producing machinery of the plants that produce tea and chocolate, the researchers discovered that *Coffea canephora* evolved a unique set of enzymes dedicated to the creation of caffeine compounds. In other words, the staple crops that deliver caffeine into the human bloodstream appear to have independently developed caffeine production, instead of descending from one common ancestor.

That convergent evolution may be partially explained by caffeine's use as a kind of chemical weapon, not unlike the pungent, burning taste of piperine. The coffee plant sheds leaves laced with caffeine, which make the surrounding soil inhospitable to other plants that might compete for resources or sunlight, and the high doses of caffeine in the berries themselves poison insects that might be otherwise inclined to eat the berries. But as the science writer Carl Zimmer has observed, the evolution of caffeine may not have been entirely defensive, noting that the coffee plant also includes low levels of caffeine in the nectar it produces. "When insects feed on caffeine-spiked nectar," Zimmer writes, "they get a beneficial buzz: they become much more likely to remember the scent of the flower. This enhanced memory may make it more likely that the insect will revisit the flower and spread its pollen further." In this one respect, our cultural exploitation of caffeine's chemistry may have run parallel with its evolutionary function, given the drug's well-studied capacity to enhance the memory centers of the brain.

From the beginning, coffee walked a fine line between medicine and recreational drug. Its utilitarian purposes—increasing short-term memory, combating drowsiness—have always been an undeniable part of the beverage's appeal. Caffeine likely played some role in the great expansion of intellectual and industrial activity that Europe witnessed in the eighteenth century, at the exact moment that coffee and tea became staples of the European diet, particularly in England and France. (The Europeans effectively swapped out a depressant—the default daytime beverage of alcohol—for a stimulant en masse, with predictable results.) Certainly caffeine was a crucial ingredient in easing the workforce of the first industrial towns into the regimented schedules of factory time. But caffeine was more than just a smart pill; an entire complex of devices and techniques evolved around the *Coffea arabica* berry to make it more pleasurable to extract the caffeine from its naturally bitter state: grinders, milk steamers, French presses, espresso machines. Tea, by comparison, never attracted the diverse arsenal of devices that coffee did, perhaps because tea was less challenging to the palate in terms of flavor. When you read the accounts of the Europeans first encountering coffee in the sixteenth century, it's hard to find a single person who enjoyed the beverage for its taste. (Those early caffeine adventurers would likely be baffled to read the descriptions on a coffee blog today.) One seventeenth-century drinker described the taste of coffee as a "syrup of soot and the essence of old shoes," while the *London Spy* called it the "bitter Mohammedan gruel."

Turning coffee into something that could be independently savored for the subtleties of its taste was a project that wouldn't be successfully achieved until the nineteenth and twentieth centuries. In the seventeenth century, what made coffee fun was as much about architecture and urbanism as it was about taste receptors. Coffee made Europeans more alert and improved their recollection, but the coffeehouse gave them a new compound, built by social connections instead of enzymes.

Sometime around 1650, a Sicilian manservant named Pasqua Rosée found himself briefly in the employ of a British merchant named Daniel Edwards in Smyrna, then part of the Ottoman Empire. Edwards had developed a taste for coffee in his travels through modern-day Turkey, and so on his return to London he brought Rosée back with him as a kind of in-house barista. Serving coffee to Edward's cronies, Rosée sensed that this still-exotic Turkish drink presented a business opportunity, and he began investigating the possibility of selling coffee in a public venue modeled after the coffee shops of Istanbul and other Levantine cities. The history at this juncture is a bit blurry: Rosée may have either gone into business with Edwards or fallen out with his employer and partnered with one of his friends. What is uncontested is that Rosée began selling coffee from a shed in a churchyard near Edwards's house in 1652, promoting it with a sign painted with an image of his own head, prompting its patrons to dub it "The Turk's Head." It was London's first coffeehouse: the origin point of an institution that would be more influential than any other public space in England over the next century and a half.

Rosée published a leaflet on the "Vertue of the COFFEE Drink," which is worth quoting at length, both for the curious pleasure of hearing someone describe an everyday object to an audience that has never experienced it, and to chuckle over how far Rosée went in evangelizing the drink's medicinal powers.

> *The Grain of Berry called Coffee, groweth upon little Trees, only in the Deserts of Arabia . . . It is a simple innocent thing, composed into a Drink, by being dryed in an Oven, and ground to Poweder,*

> *and boiled up with Spring water, and about half a pint of it to be drunk . . . It is very good to help digestion, and therefore of great use to be [consumed] about 3 or 4 o'clock afternoon, as well as in morning. [It] quickens the Spirits, and makes the Heart Lightsome . . . It suppresseth Fumes exceedingly, and there good against the Head-ach, and will very much stop any Defluxion of Rheums, that distil from the Head upon the Stock, and to prevent and help Consumption and the Cough of the Lungs. It is excellent to prevent and cure the Dropsy, Gout, and Scurvy . . . It very good to prevent Mis-carryings in Child-bearing Women.*

There's something undeniably charming about Rosée's salesmanship here; not content with the entirely accurate pitch that the "Coffee Drink" would help with digestion and headaches and provide mental stimulation, he somehow found it necessary to throw in tuberculosis, scurvy, and miscarriages as well. One can only imagine the amount of money Starbucks would make today had caffeine actually possessed all of these medicinal attributes. Anticipating the caffeinated industrial revolution that would arrive in the next century, Rosée also noted that his drink would "prevent drowsiness and make one fit for business."

Rosée's leaflet did the job, and he soon found himself selling six hundred cups—or "dishes," as they were called—in a single day. Before long, coffeehouses sprouted all across London. Within a decade, there were eighty-three coffeehouses in the city, all of them catering to men. (As Habermas notes, one of the notable differences between the London coffeehouse and its Parisian equivalent, the salon, was that women were active participants in the latter.) Excluded from the "man's world" of coffee and chatter, a group of London women published in 1674 a *Woman's Petition Against Coffee,* which thundered against the "Excessive use of that Newfangled, Abominable, Heathenish Liquor called COFFEE." (The "Calico Madam" critique that emerged thirty years later can be seen as a kind of re-

Interior of a London coffeehouse in the 17th or 18th century

turn volley.) A year after the *Woman's Petition*'s publication, Charles II joined forces in attacking the coffeehouses, fearing that they were encouraging idle behavior and seditious political movements. His "Proclamation for the Suppression of Coffee Houses" described the threat in decisive terms:

> *Where it is most apparent that the multitude of coffee houses of late years . . . and the great resort of idle and disaffected persons to them have produced very evil and dangerous effects, as well for that many tradesmen and others do herein mis-spend much of their time which might and probably would be employed in and about their lawful calling and affairs, but also for that in such houses,*

> *divers false, malitious, and scandalous reports are devised and spread abroad to the defamation of His Majestie's Government . . . His Majesty hath thought it fit and necessary that the said coffee houses be (for the future) put down and suppressed . . .*

You can hear beneath that formal syntax the guttural cry of moral panic that would echo for centuries every time new leisure spaces emerged to scandalize older generations: from the department stores of the nineteenth century, to the pool halls of the early twentieth, to the video-game arcades of the 1980s. Charles might have "thought it fit and necessary" to suppress the coffeehouses, but the citizens of London thought otherwise. After a violent outcry from both the proprietors and customers of the new establishments, Charles withdrew his proclamation only a week after proclaiming it. However useless it was as a legal intervention, the language of the proclamation made it clear that the real challenge to authority that the coffeehouse presented had little to do with the drug itself and everything to do with the social space that the coffeehouse introduced to English society. By the early 1700s, London neighborhoods featured more than a thousand coffeehouses, far more than any city in the world. (Amsterdam, at that point the only city in Europe to rival London's affluence, only had thirty-two as of 1700.) More than any other physical environment, the coffeehouse would nurture and inspire the commercial, artistic, and literary flowering of the British Enlightenment.

In part that cultural transformation was enabled by the remarkable specialization that characterized the coffeehouse ecosystem in the 1700s. Coffeehouses clustered in Exchange Alley specialized in stock-market speculation; Waghorn's in Westminster was a hotbed of political gossip. Others functioned as gambling dens or brothels; some catered to more obscure interests, as in John Hogarth's coffeehouse, where the conversation was exclusively conducted in Latin. The historian Matthew Green writes, "At the Bedford Coffeehouse

Satirical print of a "Coffee House Mob," 1710

in Covent Garden hung a 'theatrical thermometer' with temperatures ranging from 'excellent' to 'execrable,' registering the company's verdicts on the latest plays and performances, tormenting playwrights and actors on a weekly basis." Perhaps most famously, coffeehouses served as the de facto offices for a newly forming journalistic class. In the inaugural issue of the *Tatler* in 1709, Richard Steele explained how his news and commentary would be sourced from the specialized chattering classes of London's coffeehouses:

> *All accounts of gallantry, pleasure and entertainment shall be under the article of White's coffee house; poetry under that of Will's coffee house; learning under the title of Grecian; foreign and domestic news you will have from St. James' coffee house, and what else I shall on any other subject offer shall be dated from my own apartment.*

In some coffeehouses, the threads of our history of play converged. Lloyd's Coffeehouse catered to the maritime community, ultimately evolving into the insurance giant Lloyd's of London. Born in a coffeehouse, the modern insurance business used the mathematics of probability invented by dice players to make it financially viable to send ships around the world in search of fashionable new fabrics like calico and chintz.

But the cultural impact of the coffeehouses did not exclusively rely on their catering to specific niches. Many, like the London Coffeehouse, where Ben Franklin and Joseph Priestley held court in the 1760s, were intensely multidisciplinary in their interests, lacking the specialized focus of a corporate headquarters or a university department. They were spaces where intellectual networks converged. "Unexpectedly wide-ranging discussions," Green writes, "could be twined from a single conversational thread as when, at John's coffeehouse in 1715, news about the execution of a rebel Jacobite Lord (as recorded by Dudley Ryder) transmogrified into a discourse on 'the ease of death by beheading' with one participant telling of an experiment he'd conducted slicing a viper in two and watching in amazement as both ends slithered off in different directions. Was this, as some of the company conjectured, proof of the existence of two consciousnesses?"

In some influential coffeehouses that eclecticism was visible in the decor itself. In the early 1700s, a London doctor named Hans Sloane began collecting exotic items that would eventually fill nine rooms of a manor house in Chelsea. An observer in 1730 described a small subsection of Sloane's collection:

> *. . . A Swedish owl, 2 Crain Birds, a dog; vast No. Of Agats, an Owel in one, exact, orange; Tobacco in others, Lusus Naturae; an opal here; Catalogue of Books, about 40 volumes; 250 large Folios, Horti Sicci; Butterflies in Nos.; 23,000 Medals; Inscriptions, one exceeding fair from Caerleon; A fetus cut out of a Woman's belly, thought she had the dropsy; lived afterwards, and had several Children.*

By the time of his death, Sloane had assembled more than seventy thousand objects, which he bequeathed to King George II. Sloane's collection was a particularly accomplished version of what the Germans called *Wunderkammerns*—literally, "cabinet of wonders." *Wunderkammerns* were small shrines to the gods of miscellany, featuring ancient coins, trinkets, embalmed mummies, daggers, rhinoceros horns, and the like. No organizing principle united these objects; to find a home in the cabinet of wonders, objects needed only the capacity to surprise the aristocrats who were invited to survey them. But Sloane's collection was not only perused by the London elite. An enterprising barber and dentist named James Salter struck a deal with Sloane in the first years of the 1700s, borrowing a

Archduke Albert and Archduchess Isabella Visiting a Collector's Cabinet, ***circa 1621–23 by Jan Breughel the Elder and Hieronymus Francken II***

few items from Sloane's growing collection. Salter put them on display in a new coffeehouse he had opened near Chelsea Church; before long, Salter had adopted the more exotic name Don Saltero, and his cabinet of wonders and caffeine became a popular destination for London's cognoscenti, its walls and ceilings covered with hundreds of oddities: barnacles, wampum, a "flaming sword" that allegedly belonged to William the Conquerer, petrified oysters, a tooth extracted from the mouth of a giant, whips, backscratchers, and gems.

Many of Sloane's wonders originated in ancient history, but the collection they formed turned out to belong to the future. Shortly after Sloane's death, his vast archive of exotica became the foundation of the British Museum, whose charter of incorporation declared that it provide "not only for the inspection and entertainment of the learned and the curious, but for the general use and benefit of the public." It was the first national museum in history that truly belonged to the people, almost half a century before the French revolutionaries opened the doors of the Louvre.

Two very different traditions descend from Don Saltero and Sloane's curations. One line leads to the carnival huckster showmanship of Ripley's Believe It or Not: a collection of oddities of dubious origin, arranged to amuse tourists and gawkers. But the other line has a more advanced pedigree. The cabinet of wonders was the physical embodiment of a new kind of scholarship, a new model for the intellectually curious. To be a scholar in the centuries before Don Saltero was to be versed in the classics and the Bible; it implied focus and diligence and a deference to the ancient wisdoms. The intellectual imagination that flowered in the 1600s and 1700s was fundamentally different: it was multidisciplinary, global in its scope, intrigued by oddities as much as by classical knowledge. Diderot's *Encyclopédie* and the *Encyclopaedia Britannica* both emerged from the ethos of the collector. In a famous passage from *The Prelude*, Wordsworth drew upon the *Wunderkammern* as a metaphor for his desultory intellectual pursuits at Cambridge as a young man:

I gazed, roving as through a cabinet
Or wide museum (thronged with fishes, gems,
Birds, crocodiles, shells) where little can be seen
Well understood, or naturally endeared,
Yet still does every step bring something forth
That quickens, pleases, stings . . .

Wordsworth was capturing a sensibility that was enjoyed, during his day, by a tiny slice of the population, but that would become far more mainstream in the years to come: the idea of college as a time of intellectual play, a time to experiment, to dabble in eclectic interests and attitudes. The idea lives on today as a legacy of Romantics like Wordsworth, but it can be traced back even further, to the curiosity shops and coffeehouses of the Enlightenment collector.

Like the temples of illusion and department stores that would follow them, coffeehouses were environments where social classes converged: poets, lords, stock speculators, actors, gossips, entrepreneurs, scientists—all found a seat in the shared environment of the coffeehouse. It was not, to be sure, an environment that welcomed women or the working poor. (Most establishments charged a penny for admission, easily affordable to the middle class, but just dear enough to discourage common laborers.) But by the standards of the eighteenth century, it was, almost certainly, the most egalitarian room that modern Europeans had ever experienced. As early as 1665, a pamphlet on the new coffeehouse culture observed, in verse: "It reason seems that liberty / Of speech and words should be allow'd / Where men of differing judgements croud, / And that's a Coffeehouse, for where / Should men discourse so free as there?" Roughly a decade later, a set of "Rules" prescribing coffeehouse etiquette instructed, also in verse:

> *First, Gentry, Tradesmen, all are welcome hither,*
> *And may without Affront sit down Together:*
> *Pre-eminence of Place none here should Mind,*
> *But take the next fit Seat that he can find:*
> *Nor need any, if Finer Persons come,*
> *Rise up for to assign to them his Room.*

In his 1712 travelogue, John Macky noted that in coffeehouses, one regularly witnessed "Blue and Green Ribbons and Stars"—decorative emblems marking the highest echelons of British society—"sitting familiarly, and talking with the same freedom, as if they had left their Quality and Degrees of Distance at Home."

The inherent democracy of the coffeehouse was an achievement on its own, one that would play a role in political democratization over the course of the next century. But it also led to a staggering number of innovations: the first public museums, insurance corporations, formal stock exchanges, weekly magazines—all have roots in the generative soil of the coffeehouse. Countless other innovations in engineering, agriculture, and navigation were seeded by the "premiums"—or prizes—awarded by the Royal Society of Arts, founded in Rawthmell's Coffeehouse in Covent Garden in 1754. To its early critics, the coffeehouse may have looked like a space of indulgence and lethargy, a place where men went to escape what Charles II called "their lawful calling and affairs." But that escape—as scandalous as it seemed initially—turned out to be enormously productive. Escaping your lawful calling—and your official rank and status in society—not only created a new kind of leisure, it also created new ideas, ideas that couldn't emerge in the more stratified gathering places of commerce or religion or domestic life. Coffee may not have proved to be the miracle drug that Pasqua Rosée envisioned, at least in terms of the physical health of the people who drank it. But its cultural impact was nothing short of miraculous.

The century and a half that followed the first industrial breakthroughs of the 1750s is often imagined as a dark tide of factories rising across the planet, in England first, then through Northern Europe and America. But alongside that steam-powered march, a different kind of space proliferated as well, a kind of inverted image of the mechanized thunder and drudgery of the factory: spaces designed deliberately for the experience of leisure and entertainment. The coffeehouses in London, the taverns of Philadelphia, the grand department stores of Paris, the "ghost makers" in Berlin, the moving panoramas of New York—cities began to teem with new ways for people to escape "their lawful calling and affairs," for a few hours at least. Leisure, for the first time, became not so much a commodity—since part of the pleasure came from the uniqueness of each new attraction—but rather a kind of environment for sale, crafted for maximum enjoyment value.

This was a profoundly urban phenomenon, but as the illusionists and department-store magnates steadily took over the downtowns of cities across the world, a parallel revolution took place far from of the metropolitan centers. Wilderness, too, was transformed from a space of fear and hardship into a space organized around, and celebrated for, the human pleasures it induced. The romance of immersing yourself in nature, "getting away from it all," was not something that came naturally to humans, who had been walling themselves off from nature since the birth of agriculture. In 1620, when the first Pilgrims sailed into exquisite Cape Cod Bay, passing the dunes outside of Provincetown that would centuries later attract vacationers from around the world, they reported that they had found themselves in "a hideous and desolate wilderness full of wild beasts and wild men." Scenes that today evoke grandeur and awe were in many cases abhorrent to a seventeenth-century eye, at least among the Europeans who were educated enough to record their impressions. Mountains in particular were thought to be aesthetically offensive. They were called "warts," "boils," and even in one bizarre case "nature's pudenda." As late as the eighteenth century, travelers through the Alps would often ask to be blindfolded to avoid looking at the awful scenery. When they looked at a mountain, they imagined it as a habitat they might have to survive in, rather than as a postcard image of Natural Beauty. While human beings seem to have an innate fondness for natural greenery—what E. O. Wilson famously called biophilia—that deep-rooted instinct appears to have been overpowered by ten thousand years of agricultural and urban settlements. Wilderness back then was something to conquer, not contemplate.

Horace-Bénédict de Saussure

The word *innovation* is conventionally used to describe advances in science or technology; the lightbulb is an innovation, as is the iPod. But our lives today have been shaped not only by new gadgets but also by new ideas and attitudes that had to be nurtured and disseminated by people who were in many cases as visionary and as driven as Thomas Edison or Steve Jobs. The idea of nature as a space to seek out and savor aesthetically was one of those attitudes. It seems intuitive to us today, living in a world where national parks attract tens of millions of visitors a year. But that sensibility was itself a kind of cultural innovation, one that that came together in the late 1700s and early 1800s.

If the very idea of the "great outdoors" was an innovation, then the question becomes: Who were the innovators behind it? The conventional story points to the poets and painters—to Wordsworth and Keats and Turner—who began explicitly thinking of nature as the phenomenon most likely to trigger that mix of awe, pleasure, and fear that the Romantics called "the sublime." But our modern sense of nature as a space to be enjoyed as a recreational act also dates back to an eighteenth-century scientist named Horace-Bénédict de Saussure. Born in Geneva in 1740, de Saussure was a member of the Swiss aristocracy, but he was also a man of science: a botanist, geologist, and a natural philosopher. De Saussure was particularly fascinated by those hideous monstrosities of the natural world—mountains—believing that they offered tantalizing clues about the geological composition of the earth and its atmosphere. But turning those clues into genuine understanding meant the mountains would have to be climbed.

One mountain in particular caught de Saussure's attention: Mont Blanc, the highest peak in Western Europe, popularly known as "the Accursed Mountain." De Saussure believed that if he could get to the top, he could collect valuable data about the earth and its atmosphere. But mountaineering was effectively unknown as a skill, precisely because summits like Mont Blanc offered nothing useful or valuable to the humans gazing up at their peaks. Humans had navigated oceans, built canals, crossed deserts, but always because some reward (real or imagined) lay at the end of the journey. Climbing fifteen thousand feet of ice, snow, and rock with nothing to reward you but a sense of achievement made no sense—particularly when the mountains were rumored to be the habitat of monstrous creatures. As late as 1723 a Swiss fellow of the Royal Society published a detailed description of the dragons that lived in the Alps.

De Saussure recognized that he didn't have the skills and fortitude to discover a route to the top of Mont Blanc on his own, and so he offered a reward to the first climber to make the ascent. On August 8, 1786, the French climbers Jacques Balmat and Michel Paccard reached the summit for the first time, and claimed de Saussure's reward shortly thereafter. With the route now mapped, de Saussure set out to become one of the first to follow in the footsteps of Balmat and Paccard. As a well-bred aristocrat, he made the historic ascent in style, with a team of eighteen servants and guides, and a kit that included an entire bed, along with "mattresses, sheets, coverlet and a green curtain," a tent, a ladder, a parasol, two frock coats, two nightshirts, two cravats, and a pair of slippers.

The cravats and coverlets may seem a little rich to the modern climber, used to the extreme efficiency of outdoor gear, but de Saussure was not just bringing the niceties to the summit of Mont Blanc. He also brought an extensive supply of scientific tools to capture as much data as possible from the peak. He used two large barometers to confirm the height of the summit; he measured the air temperature and the humidity and recorded the boiling point of water at altitude. He fired a pistol to record the effect of altitude on sound, and measured the pulses of his companions, taking detailed notes on their sense of smell and taste at high altitude. He recorded the stratigraphy of rock types and tracked which flowering plants and animal life were able to survive at such a high elevation—including two butterflies flying above the snowline. He even measured the precise color of the sky, using a custom color chart of his own design called a cyanometer. According to his analysis, the sky was the deepest blue he'd ever seen: "39 degrees blue."

The science was groundbreaking on multiple levels. Perhaps most significantly, de Saussure's analysis of mountain landscapes suggested to him that the earth was much older than had been conventionally assumed, evidence that would become a cornerstone of Darwin's theory of evolution a century later. But de Saussure's description of the view ended up being as influential as his barometric readings and geological surveys. He later published a book recounting his mountaineering adventures. "I could hardly believe my eyes; it appeared to me like a dream, when I saw placed under my feet those majestic summits," he wrote of Mont Blanc. For some reason, the words resonated in the public imagination. Many followed in de Saussure's footsteps to see the incredible sight for themselves, marking the birth of alpine tourism. A small industry in Mont Blanc memorabilia quickly sprouted; miniature scale models of the mountain were sold to tourists. De Saussure donated bits of granite he collected from the summit to prominent institutions for display and study. For a while, owning a fragment of Mont Blanc was like owning a piece of NASA-certified moon rock or a pebble from the Berlin Wall.

De Saussure's explorations—and his travelogues—helped transform the general public's relationship to nature. By the end of the century, the rhetoric describing mountain landscapes had been entirely reversed. No longer warts and boils, alpine summits became "palaces of nature" and a "terrestrial paradise." But mountains were only part of the transformation. A rising middle class began looking at nature with new eyes: remote mountains, valleys, canyons, waterfalls, lakes, and rivers became marquee destinations. Most unusual ecosystems or geographical features were entirely unknown to the

A Claude glass, 18th century

average citizen at the dawn of the eighteenth century, and even the educated classes would never think of setting off to visit a mountain or ocean-side cliff just for the aesthetics of the experience. But de Saussure and his followers changed all that. Nature tourism emerged as a leisure pastime. No longer a space to be feared, nature itself became a kind of wonderland, protected and mapped for the purposes of human delight. New gadgets arrived to enhance the tourist experience as it was happening, ironically by offering a degraded version of the spectacle they had traveled to see. Standing at the base of the Matterhorn or gazing out over the cliffs of Gibraltar, tourists captured the vista with an odd contraption called a "Claude glass," named after the landscape painter Claude Lorrain. The device was effectively a tinted mirror, designed to re-create the aesthetics of an oil painting, not unlike the deliberately degraded filters of Instagram. Tourists turned their back on the mountain or the waterfall, held up the tinted mirror, and appreciated the vista in the reflection.

The shifting relationship to nature was reflected, and amplified, by changes in artistic expression. Dramatic landscapes shifted from the backgrounds to the foregrounds of paintings. In the United States, paintings and early photographs of the American west were deployed by the railroads to encourage migration westward. Some of this new obsession with nature was stoked by the illusionists, who brought thrilling virtual renditions of wilderness to metropolitan centers. In addition to John Banvard's *Grand Moving Painting of the Mississippi and Missouri Rivers*, moving panoramas which also offered tours that allowed spectators to simulate a ride down the Nile or a journey across the epic landscapes of the American west to California and Oregon. Mountaineering, too, found a place in the palaces of illusion. In the 1850s, a British showman and mountain climber named Albert Smith created a moving panorama that documented his own alpine conquests. Smith built a simulated Swiss chalet inside London's Egyptian Theater, where rapt audiences watched the panorama unfurl as Smith narrated his exploits. The show—titled *Albert Smith's Ascent of Mont Blanc*—played more than two thousand times to sold-out crowds during the 1850s.

Inevitably, as more people fell in love with untouched wilderness, a growing chorus began to argue for preserving these spaces indefinitely. In April of 1872, an act of Congress declared a block of land in the territories of Wyoming and Montana "as a public park or pleasuring-ground for the benefit and enjoyment of the people," the first national park in human history. Today, Yellowstone National Park attracts more than three million visitors a year, and 1,200 national parks of comparable scale have been established in more than a hundred countries around the world.

New cultural institutions often emerge at the intersection points of older strains, like a complex of streams that converges to form a single river. (Think of the way the cinema arose out of the nexus of the magic-lantern illusionists, early photographers, traditional theater, and the thaumatrope.) By the end of the nineteenth century, a new kind of wonderland became imaginable, one that took the cosmopolitan ethos that had been growing for the preceding two centuries and turned it into a weekend attraction. The wellsprings that fed this new form were multiple: the new interest in nature as spectacle; the runaway popularity of Great Exhibitions, like the Crystal Palace of 1851, that showcased objects of wonder from around the world; the new urban parks being designed in New York and Paris and Boston; and the roving circuses of Barnum and Bailey. Many of these environments derived from conventions that had first been developed behind the fences of royal estates and other aristocratic properties: follies, gardens—nature sculpted and arranged for the amusement of an idle stroll or a carriage ride.

The public zoos that first appeared in European and American cities in the middle of the nineteenth century embodied these many influences: the newfound interest in experiencing nature; the global vistas of an imperial age; the private menageries of royal courts now opened to the public. The London Zoo opened in Regents Park in 1828 as a purely scientific institution, but quickly began a practice of admitting well-to-do members of society who wished to enjoy its exotic collection of animals. By the 1840s, the zoo had opened its doors to the general public. Its benefactors believed that the pastime of viewing lions, elephants, and rhinos would function as what they called "rational recreation," delighting the mind as well as the body and widening one's appreciation for the natural bounty of life on earth—all in a few acres in the middle of a teeming metropolis. A guidebook to the zoo from the 1840s outlined what the average zoo visitor might expect:

> *In his mind's eye he may track the pathless desert and sandy waste; he may climb amid the romantic solitudes, the towering peaks, the wilder crags of the Himalayan heights, and wander through the green vales of that lofty range whose lowest depths are higher than the summits of European mountains; or he may peer among the dark lagoons of the African rivers, enshrouded by forests whose rank green foliage excludes the rays of even a tropical sun; for here he has the evidence and the fruits of those countries which have hitherto been only to him an impalpable and uncertain idea.*

The dream of global experience that had begun with the spice trade's "taste of the Orient" had taken on a new material reality: you could now see (and no doubt smell) creatures from the four corners of the world without even leaving the city limits. The "impalpable and uncertain idea" of creatures from Africa or Asia was now, thrillingly, staring straight into your eyes.

One of those creatures was an orangutan named Jenny, who had been brought to the zoo in 1837 after the park's original orangutan, Tommy, died of tuberculosis after only a few months in captivity two years earlier. Stories of manlike apes that lived in Africa had fascinated Europeans for centuries, and a handful of skeletons and specimens mounted by taxidermists had appeared in museums and *Wunderkammerns*; but until Jenny and Tommy's arrival, none of the great apes had survived a sea voyage to England. The public's interest in Jenny was naturally quite expansive, an interest that was stoked by the zoo's insistence on dressing both Tommy and Jenny up in human clothes. Tommy sported a "Guernsey frock and a little sailor's hat," while Jenny was dressed like a proper British schoolgirl. The apes were taught to take tea, and Jenny eventually learned to follow many spoken commands from the zookeepers. Queen Victoria and Prince Albert witnessed one of the orangutans enjoying tea in 1842; she wrote in her diary that she was "too wonderful," but also "frightful, and painfully and disagreeably human."

The sight of an orangutan in fancy dress sipping a cup of Earl Grey may seem more *Bedtime for Bonzo* than something you might have expected from a serious zoological society, but Jenny's fashion sense and manners played a critical role in the nineteenth century's greatest scientific breakthrough. In the spring of 1838, Charles Darwin spent several rapt hours watching Jenny, "in great perfection." During his stay, the keeper showed Jenny an apple, but then withheld it, and the two began a complex exchange that astounded Darwin. He later recalled:

> *She threw herself on her back, kicked and cried, precisely like a naughty child. She then looked very sulky and after two or three fits of passion, the keeper said 'Jenny if you will stop bawling and be a good girl, I will give you the apple.' She certainly understood every word of this, and though, like a child, she had great work to stop whining, she at last succeeded, and then got the apple, with which she jumped into an arm chair and began eating it, with the most contented countenance imaginable.*

Several weeks after his visit to the zoo, but months *before* the famous "Malthusian epiphany" where he allegedly hit upon the theory of natural selection, Darwin wrote a passage in his journal that dared to suggest the evolutionary link that would so shock the world more than two decades later. "Let man visit Ourang-outang in domestication, hear its expressive whine; see its intelligence when spoken [to], as if it understood every word said; see its affection to those it knew; see its passion and rage, sulkiness and very actions of despair; let him look at the savage . . . and then let him dare to boast of his proud pre-eminence . . . Man in his arrogance thinks himself a great work, worthy the interposition of a deity. More humble and I believe true to consider him created from animals." We think of natural selection as an idea that required a voyage across the world, far from London, to the Galápagos and beyond. But the most controversial element of Darwin's idea arose in part because the world had come to London, in the "rational recreation" of the Regents Park Zoo.

The impact of the nineteenth-century zoo would not always be quite so erudite. In the years after Darwin's visit with Jenny, the growing interest in nature as a kind of entertainment experience fed a blossoming of public zoos across Europe; in the 1860s, a new zoo opened on average every year in Germany. With the general public no longer satisfied by the stuffed animals on display in museum dioramas, a new class of wild-animal traders emerged, supplying zoos, circuses, pet shops, and menageries around the world with thousands of exotic animals. A cross between P. T. Barnum and Indiana Jones, a German animal trader named Carl Hagenbeck became a minor celebrity, with a reputation for daring adventures to the far corners of the earth, capturing dangerous beasts and bringing them back alive to the urban centers of Europe. (The character of Carl Denham in *King Kong* was cast from the mold that Hagenbeck first created.) Hagenbeck truly was an inveterate traveler, averaging more than thirty thousand miles in a year—the equivalent of ten flights across the Atlantic. In the age of trains and steamships, that was a staggering amount of time on the road. But Hagenbeck did almost none of the actual animal capture himself; instead, he ran a vertically integrated system that stretched from trappers in sub-Saharan Africa to the showrooms and expositions that Hagenbeck began establishing across Europe and the United States. *Wild-animal trader* has an undeniably buff ring to it as a job description, but in the end Hagenbeck's success was largely due to his skills at supply chain management. Notoriously, the showman did not limit himself to the animals of the world; he also staged a number of hugely successful—and, to modern eyes, hugely offensive—exhibitions of "savages in their natural state": Inuits from Labrador, Nubians from the Egyptian Sudan.

Yet, for all his travels, Hagenbeck's most important legacy didn't take shape until he settled down. In 1897, he bought a thirty-five-acre estate in Stellingen, outside Hamburg, and set out to build a new kind of park. "An enormous task lay before us: to transform a wasteland into a luxury park," he later wrote, "a pleasure park with waterfalls and mountain formations, with practical animal shelters and buildings devoted to pure leisure." His eventual creation, Tierpark Hagenbeck, is now considered one of the world's first theme parks. Carnivals and traveling fairs had existed for centuries, of course; mechanical rides had entertained visitors at the 1893 Chicago Columbia Exposition; Hagenbeck himself had contributed a flagship attraction to Coney Island's Luna Park when it opened in 1903. But no one before him had ever dared to build a permanent park as a kind of simulated world, a fully immersive experience that created the sense of global exploration and adventure without any of the actual risks that genuine exploration required.

Opening in 1907, Hagenbeck's park would eventually stretch over eighty acres; anticipating Disney's designs, he created fake mountain landscapes that concealed animal shelters, service entrances, heating pipes, and an electric motor that powered a waterfall. Across his artificial landscape Hagenbeck placed animals, native people, and villages: in a sense, an enormous moving panorama populated by live beings. A visitor entering the park would see a pond of aquatic birds with a field of deer beyond, stretching up to a ravine inhabited by lions, all backed by a mountain ridge beyond which ibex and wild sheep grazed. Paths, tunnels, and trails wound through the faux terrain with vistas at the peaks, where tourists could pretend to be de Saussure contemplating the view from the summit of Mont Blanc. Instead of bars or glass, Hagenbeck used the terrain to separate predators from prey, building deep but invisible trenches to keep the spectators and native people safe. (Thomas Edison visited the park and was apparently terrified when he walked around a group of trees and

Tierpark Hagenbeck, one of the world's first theme parks

came face-to-face with a lion.) Some of the attractions combined immersive landscapes, zoological display, and rides. At the summit of the "Northern Plateau," visitors could survey a frozen expanse of a simulated Arctic Sea populated by polar bears and seals, and then board a "Canadian tobogganing slide" that whisked them through a rocky crevasse, past the polar bears and a herd of caribou.

The park was an immediate hit: more than ten thousand turned up to its opening day; by 1914, 7.5 million people had passed through its gates. "When you approach the Lions' Ravine," one visitor recalled, "surrounded and vaulted by majestic cliffs, you experience the feeling as though you were walking in the desert and happened upon a brook of lions, and, since you know that you are safe, you enjoy this sensation as purely aesthetic." The mix of innovations that made that "pure aestheticism" possible had an undeniable commercial impact on the twentieth century. The theme-park industry today is one of the most lucrative entertainment businesses in the world. (In the United States alone, amusement parks generate more than $50 billion in economic activity each year.) But these fantasylands turned out to have philosophical implications as well. A rich tradition of continental philosophy emerged in the 1970s—most famously Umberto Eco's *Travels in Hyperreality* and Jean Baudrillard's *Simulacra and Simulation*—decrying the illusory artifice of modern culture, all the theme restaurants and megamalls and old downtowns converted into spectacles of consumption. "Disneyland is presented as imaginary in order to make us believe that the rest is real," Baudrillard famously announced in *Simulacra and Simulation*, "whereas all of Los Angeles and the America that surrounds it are no longer real, but belong to the hyperreal order and to the order of simulation." Written by Europeans who represented an old-world philosophical tradition, albeit ones with an interest in pop cultural semiotics, Eco and Baudrillard's travelogues contained a healthy measure of anti-American condescension, as though the escapism of Disney's fantasy parks was a kind of ideological virus that the United States had unleashed on the world. But Carl Hagenbeck's Trierpark is a useful corrective to that story: theme parks hit their mature phase in postwar America, but they spent most of their adolescence in Europe, among the ghost makers and illusionists and wild-animal traders.

Hagenback may have propelled us toward Eco's hyperreal future with his simulated mountain ranges and fake savannas. But in the end, his creation may have turned out to be all too real. He died in 1914; legend has it that he was bitten by a venomous tree snake that he had brought back from Africa.

The modern world is now brimming with "pleasuring grounds," in cities, suburbs, and rural parts of the globe. The market capitalization of Starbucks is currently $85 billion, making it one of the most valuable restaurant and food-service companies in the world. Where cities once nurtured the palaces of illusion and other entertainment venues, now the escapist fantasies nurture entire cities. Orlando, regularly the most visited city in the United States, is an urban center almost entirely created by the decision of Walt Disney to plant his theme park there. Las Vegas, a town that had a population of ninety-six people just a century ago, has been the fastest-growing city in the United States over the past decade.

It is easy—and probably not wrong—to be a cynic about the pleasuring grounds of the twenty-first century, at least in the United States and Europe. The revolutionary sentiments that took root at the Green Dragon in Boston are not likely to be stirring at the sports bars that line Fenway; it's hard to have a multidisciplinary coffeehouse salon when everyone in Starbucks is staring at a laptop, wearing headphones. (The second half of Habermas's analysis of the public sphere was dedicated to its modern-day demise.) But at least one part of this legacy continues to inspire: the urban parks that were first built in the second half of the nineteenth century, as an outgrowth of our new relationship to nature that de Saussure and the Romantics fostered, and as a reaction to the runaway growth and overcrowding of the new metropolitan centers. These spaces continue to be vital connection points for the many cultural threads that create a great city. Consider Brooklyn's Prospect Park, designed by Frederick Law Olmsted and Calvert Vaux in the 1860s, after the completion of Manhattan's Central Park. Walk through the barbecue and picnic areas on a Fourth of July afternoon and it is impossible not to be moved by the exceptional diversity of the city on display: the huge clan of Korean-Americans gathered under an elm tree,

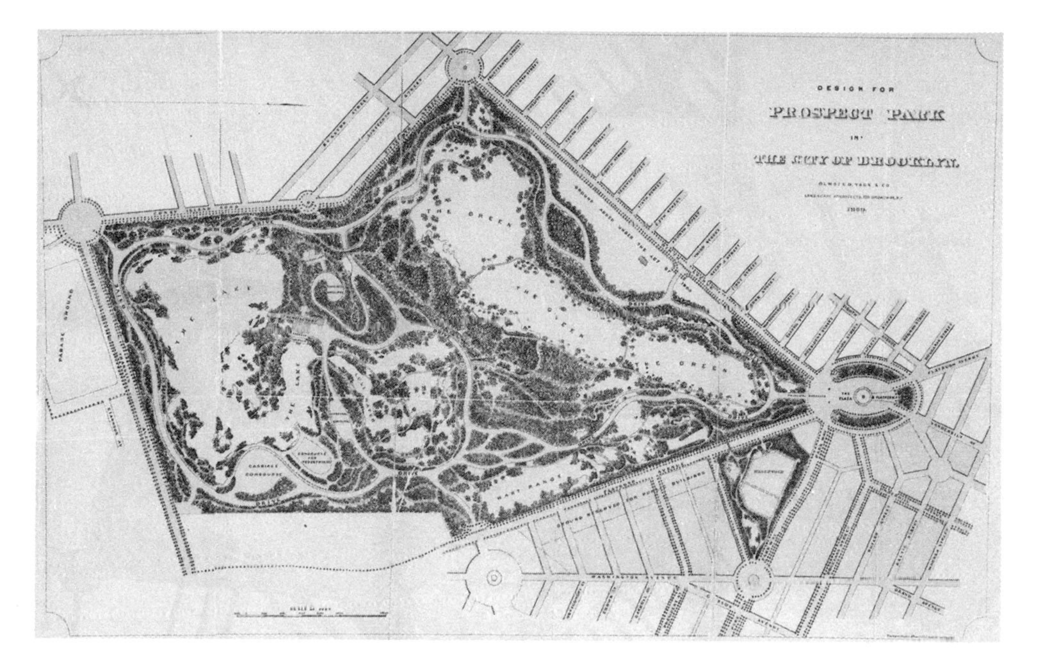

Design for Brooklyn's Prospect Park, circa 1868

with the family of Hasidic Jews strolling down the path behind them; the Puerto Ricans barbecuing up the hill, with the Williamsburg hipsters playing Frisbee between them; the rap and salsa and acoustic guitars; the old couple reading Spanish-language papers on a park bench. There is an entire world hanging out together in a space the size of ten city blocks, and the space is as safe and green and at ease with itself as one could possibly imagine.

No doubt such a scene would have appalled the authorities that condemned John Hughson to death for daring to create establishments where whites and blacks could enjoy their leisure time together; no doubt Charles II would only see "idle and disaffected persons" escaping their "lawful calling and affairs." But most of us today can appreciate that holiday scene for the extraordinary achievement it is. Once you get past the Macy's fireworks display, Fourth of July imagery and rhetoric is usually full of old-time Americana: the small town's one fire truck decked out for the main-street parade, the Little League game, the white picket fences with their patriotic bunting. There is plenty to celebrate about the joys of small communities, but in a way, there is nothing particularly original about that story. World history is teeming with small, successful communities united by a common culture and worldview, after all. What is much rarer is that Fourth of July scene in Prospect Park, and in most urban parks in metropolitan centers around the world. Until modern times, social spaces where all those different groups could happily coexist were almost unheard-of. Today, we take these environments for granted, and rarely celebrate the visionaries that helped bring them into being. But in a way, they are an achievement every bit as impressive as the conventional icons of modern progress: skyscrapers or cell phones or satellites. You can see them as the completion of a project that began thousands of years before, with the global network that first took shape around the spice trade. The taste of clove and nutmeg and pepper compelled human beings to explore the planet and erect markets that stretched to all corners of the globe. Today, that connected world is just down the street from us, reading the paper on a park bench or grilling hot dogs: at peace with itself, and at play.

Conclusion

The Surprise Instinct

For many years the conviction has grown upon me
that civilization arises and unfolds in and as play.

—JOHAN HUIZINGA,
Homo Ludens

A history of play and wonder would be justified even if it focused exclusively on the pleasure those experiences have brought us. The fact that our lives are surrounded by institutions designed with the specific purpose of bringing us happiness and amusement has to count as one of the undeniable achievements of civilization, however "uncivilized" many of those amusements first seemed. (And still seem, to some.) It would be entirely sufficient to write a history that just tracked the "pleasuring-grounds" and toys on their own merits. The world is a more interesting place because there are coffeehouses and national parks and IMAX theaters in it; we should celebrate the people behind those institutions the way we celebrate and study high-tech innovators or political revolutionaries.

Yet, still: Turn your mind's eye away from all those wonderlands of illusion and delight and think about the utilitarian story. Ignore the pleasure those institutions generated, and focus on the innovations or historical sea changes they helped bring about: public museums, the age of exploration, the rubber industry, stock markets, programmable computers, the industrial revolution, robots, the public sphere, global trade, probability-based insurance policies, the American Revolution, clinical drug trials, the LGBT rights movement, celebrity culture. Think, too, of all the tragic consequences that descend from our endless quest for delight: slavery and exploitation and conquest. The sheer magnitude of this influence is remarkable. How odd it is that slacking off on one's "lawful calling and affairs" would set off so many commercial and scientific aftershocks. The pleasure of play is understandable. The *productivity* of play is harder to explain.

Making sense of this mystery requires that we peer into the inner workings of the human brain, drawing on recent research in neuroscience and cognitive psychology—research that, fittingly, began by studying games. In the 1950s, inspired by Alan Turing's musing on a chess-playing computer, a computer scientist at IBM named Arthur Samuels created a software program that could play checkers at a reasonable skill level on an IBM 701. (Legend has it that when IBM CEO Thomas Watson saw an early draft of the program, he predicted that the news of the checkers game would cause IBM stock to jump fifteen points.) As his work developed, Samuels grew increasingly interested in not simply teaching the computer how to play checkers but in having the computer learn on its own, through experience. This line of inquiry led to self-learning algorithms for more complicated games like chess and backgammon developed in the 1960s and 1970s, but more importantly, it led to a model of learning that has come to shape our understanding of the human mind itself. The model has several variants, each with its own adherents and detractors; it includes theories called "temporal difference learning," or the "Rescorla-Wagner model," or "reward prediction error." Beneath these distinctions, though, the model suggests a common principle: humans—and other organisms—evolved neural mechanisms that promote learning when they have experiences that confound their expectations. When the world surprises us with something, our brains are wired to pay attention.

The early checkers and backgammon applications relied on this principle to bootstrap themselves into a high level of play. The software would begin with a rough model of what successful strategy looked like, making predictions about the consequences of each of its moves. Over time, it learned by paying careful attention to the *difference* between its predictions and the actual outcome. Based on those constructive errors, the software would then alter its model for the next game; after thousands of iterations, the software learned a high-level strategy without any expert player advising it directly. In a way, the AI researchers had programmed an appetite for surprise into the software.

Psychologists have long understood that this appetite is an integral part of the human mind. Countless studies of newborn infants have shown that before we can crawl or grasp or communicate, we seek out surprising phenomena in our environment. But it wasn't until the 1990s that scientists first recognized that the surprise instinct is heavily regulated by the neurotransmitter dopamine. Because drugs like cocaine and nicotine activate the dopamine system as well, popular accounts of the neurotransmitter often make the mistake of referring to it as the brain's "pleasure drug." But this shorthand description is misleading: dopamine on its own doesn't trigger feelings of pleasure the way, for instance, endorphins do. Rather, dopamine seems to help steer the attention and motivation systems of the brain. A new theory proposes that dopamine release creates a "novelty bonus" that accompanies the perception of some new phenomenon or fact about the external world. By heightening your mental faculties, making you more alert and engaged, the "novelty bonus" encourages you to learn from new experiences. (The computer scientist Jürgen Schmidhuber developed a similar process for machine learning that used a "curiosity reward" that encouraged the software to explore data with surprising results, and ignore predictable regions.) The surge of dopamine that accompanies a novel event sends out a kind of internal alarm in your mind that says: *Pay attention. Something interesting is happening here.*

Bone flutes, coffee, pepper, the Panorama, calico, Babbage's dancer, dice games, the Bon Marché—beneath all the surface differences between these objects, one common characteristic unites them all: they were *surprising* when they first appeared. We were drawn to them compulsively because they offered novel experiences, tastes, textures, sounds. Illusions took our visual predictions about the spatial arrangement of objects in the world and confounded those expectations in startling ways. Spices offered exotic new flavors that our tongues had never experienced before. One of the defining characteristics of games—as opposed to, say, narrative—is precisely the fact that they turn out differently every time we play them; games are novelty machines. That's what makes them fun (and sometimes addictive). All these forms of escape and amusement provided a "novelty bonus" to the brains that first experienced them.

On the one hand, our understanding of the dopamine system helps us understand why human beings became so obsessed with seemingly frivolous things like nutmeg or the Phantasmagoria. It is in our nature to seek out things that surprise us. But the "surprise instinct" also helps us answer a more complicated riddle: the innovative power of play, the way in which play compelled us to new cultural institutions that had little to do with our biological drives. A long tradition exists of intellectuals butting heads over the boundaries of nature and nurture; some scientist proposes a biological instinct for some facet of human behavior and inevitably a counterforce of humanities professors argues that the behavior is rooted in cultural adaptations, not some genetic destiny. But an appetite for surprise complicates those easy oppositions. Genes tend to steer us toward predictable goals, or away from predictable threats: seek out energy-rich carbohydrates; avoid intense cold or heat; find a mate and sexually reproduce with him or her. This is one reason why Darwinian interpretations of society or art tend to be less enlightening: you don't need to be an evolutionary biologist to understand that people like to fall in love or care for their children. In a way, those genetic drives are conservative in their effects. They steer us back to predictable patterns: family, shelter, food.

But the surprise instinct propels us in the opposite direction. Its object is by definition undefined. It rewards you not for finding a mate or bonding with your child or consuming energy-rich food—it rewards you for having a new experience. It rewards you for breaking out of your usual habits, for stumbling across something that confounds your expectations. This appetite has an inevitable tendency for expansion. Hunger or the need for social bonding can be satisfied by a reliable source of food or close friends. But surprise requires new blood: one generation's miracle is the next generation's old news. (As we have seen, that expansion was often geographical as well as conceptual; we live in a global economy today in large part because of the novelty bonus of pepper, cotton, and coffee.) Sometimes cultural change happens because important ideas build on each other, one insight unlocking the door to further insights. Sometimes change happens out of necessity, out of the drive to satisfy our basic survival needs. But just as often cultural change happens because human beings are bored with the old experiences, and have a hunger for something new. This is the strange paradox of play and its capacity for innovation: play leads us away from our instincts and nature in part *because* of our instincts and nature.

Because new things are strange and not immediately applicable to life's most pressing issues, they are not taken seriously. But we underestimate their ultimate significance at our peril. The drive for novelty puts us into unexpected situations, or exposes us to new materials: taverns and coffeehouses, rubber balls and magic lanterns.

Once exposed, we end up using those spaces and those devices as platforms for the ideas and revolutions of traditional history. Toys and games, as Charles Eames said, are the prelude to serious ideas. So many of the wonderlands of history offered a glimpse of future developments because those were the spaces where the new found its way into everyday life: first as an escape from our "lawful calling and affairs," and then as a key element in those affairs.

Think back to Charles Babbage, staring into the eyes of that automated doll in Merlin's attic, two centuries ago. That encounter was, quite literally, child's play, but the ideas and technologies that were stirring beneath the surface of that meeting are still transforming society as I write. Today we worry about dystopian futures where the machines become so physically dexterous that they take over our manufacturing workforce, or so intelligent that they become our masters. But perhaps, knowing the history, we have been focused on the wrong fears. Perhaps we have been wrong to worry about what will happen when the machines start thinking for themselves. What we should be *really* worried about is what will happen when they start to play.

ACKNOWLEDGMENTS

From the beginning I've thought of this book as the second in a longer series on the history of innovation, a series that began with *How We Got to Now*. So I should begin with a heartfelt thanks to the incredibly playful team that helped dream up that multi-platform project, starting with the visionary Jane Root. (Years ago Jane suggested the phrase "clever pleasure" to me to describe the kind of television we could make, a phrase that could have been an alternate title for this book.) At Nutopia, thanks to Peter Lovering, Helena Tait, Carl Griffin, Sophie Mautner, Jemima Stratton, Fleur Bone, and Jessica Cobb. (And in the extended Nutopia family, Neil Sieling.) I'm especially grateful for the research help early on from Jemila Twinch and Fred Hepburn. Matt Locke, Ian Steadman, and the rest of the How We Got to Next team have supported my work and research in countless ways over the past two years.

At PBS and WETA, thanks to Beth Hoppe and Bill Gardner; at the Lemelson Foundation, Carol Dahl, Tak Kendrick, and David Coronado; and at the Gates Foundation, Dan Brown and Miguel Castro. I first began mulling the idea of a book about the cultural innovations of play while shooting the "Sound" episode of *How We Got to Now*, which was the most *Wonderland*-like episode of that season. Thanks to Julian Jones for supporting those early ideas, and for not getting me killed in the Anza-Borrego desert.

This marks the tenth book that I have collaborated on with my agent-for-life Lydia Wills, whose contributions continue to be invaluable on many levels. It's also my seventh book with the dream team at Riverhead: Geoff Kloske, Katie Freeman, Kate Stark, Kevin Murphy, and Hal Fessenden. My brilliant new editor, Courtney Young, widened the scope of this book—and its cast of characters—in many significant ways. And I'm also very grateful to Helen Yentus and Ben Denzer for what may well be my favorite jacket design of all of my books.

A number of people were gracious enough to read the book (or sections of it) in draft form. I'm deeply indebted to the comments, corrections, and encouraging words from Alex Ross, Ken Goldberg, Stewart Brand, Steven Pinker, Mike Gazzaniga, Filipe Castro, Jane Root, Fred Hepburn, Chris Anderson, Juliet Blake, Angela Cheng, and Jay Haynes. As always, my wife, Alexa Robinson, read every word—but only improved every *other* word—with her wisdom and line-editing mojo. Thanks to Franco Moretti for introducing me to the kleptomaniacs of Paris more than two decades ago. And thanks to Jay Haynes, Annie Keating, Alex Ross, and Eric Liftin for so many conversations about music and the mind over the years.

Finally, a word of gratitude to my sons—Clay, Rowan, and Dean—for keeping me in touch with the gaming world, from Minecraft to H1Z1, from Kingdom Builder to Far Cry. I love and respect the energy and creative spirit that you bring to your life in games. Now it's time to turn off the computer and go read a book.

July 2016
Marin County, California

NOTES

Introduction

1 **"Every household was plentifully supplied":** William Stearns Davis, ed., *Readings in Ancient History: Illustrative Extracts from the Sources*, 2 vols. (Boston: Allyn and Bacon, 1912–1913), Vol. II: *Rome and the West*, 365–67.

3 **Al-Mansur founded a palace library:** Jonathan Lyons, *The House of Wisdom: How the Arabs Transformed Western Civilization* (New York: Bloomsbury Publishing, Kindle edition), 52.

6 **"Paradise Translated and Restored":** Richard Daniel Altick, *The Shows of London* (Cambridge, MA: Harvard University Press, 1978), 56.

7 **Jaquet-Droz's son, Henri Louis:** Ibid., 63.

8 **"Ladies and Gentlemen":** "John Joseph Merlin—Part One," *Georgian Gentlemen*, March 11, 2013, http://mikerendell.com/john-joseph-merlin-part-one/.

8 **"prowled the borderlines":** Simon Schaffer, "Babbage's Dancer and the Impresarios of Mechanism," *Cultural Babbage: Technology, Time, and Invention*, Francis Spufford and Jenny Uglow, eds. (London: Faber & Faber, 1996): 54.

9 **"The motions of her limbs":** Charles Babbage, *Passages from the Life of a Philosopher* (Cambridge University Press, 2011), 32.

12 **"the hummingbird effect":** Steven Johnson, *How We Got to Now: Six Innovations that Made the Modern World* (New York: Riverhead, 2014), 4–7.

14 **"It may sometimes happen":** Samuel Johnson, *Rambler*, no. 8.

Chapter 1. Fashion and Shopping

18 **"The Phoenicians' now-proven aptitude":** Simon Winchester, *Atlantic: Great Sea Battles, Heroic Discoveries, Titanic Storms, and a Vast Ocean of a Million Stories* (New York: Harper, 2010), 68.

20 **And yet the archeological record:** "By c. 36,000 to 28,000 BCE grinding, shaping, and polishing allowed Neolithic 'jewelers' to produce beads in the shape of female breasts and torsos, while others benefited from advances in ceramic technology and created miniature clay-fired animal figurines that they mounted on cords. Archeologists often find beads in burial sites of the Upper Paleolithic and Neolithic

Periods." Phyllis G. Tortora, *Dress, Fashion and Technology: From Prehistory to the Present (Dress, Body, Culture)* (New York: Bloomsbury Publishing, Kindle edition), Kindle locations 659–663.

22 **"The seductive design of shops":** Claire Walsh, "Shop Design and the Display of Goods in Eighteenth-Century London," *Journal of Design History* 8.3 (1995), 162.

24 **"a kind of enchantment which blinds":** Walsh, 171.

24 **"Painting and adorning":** Walsh, 163.

25 **"This afternoon some ladies":** Walsh, 171.

28 **"a taudry Callico Madam":** Chloe Wigston Smith, "Calico Madams: Servants, Consumption, and the Calico Crisis," *Eighteenth-Century Life* 31, no. 2 (2007), 32–33.

30 **"The spectacular early triumph":** John Styles, *The Dress of the People: Everyday Fashion in Eighteenth-Century England* (New Haven, CT: Yale University Press, 2007), 321.

30 **"It is not Necessity":** Nicholas Barbon, *A Discourse of Trade*, Printed by Tho. Milbourn for the author (London: 1905), 35.

32 **But the Calico Madams suggest**: McKendrick et al. pose the question clearly, though they are themselves still working under the convention that the consumer revolution was largely the work of men: "Some discussion is required of why attention centred on the great industrialists and the supply side of the supply-demand equation, and why so little attention has been given to the hordes of little men who helped to boost the demand side and who succeeded in exiting new wants, in making available new goods, and in satisfying a new consumer market of unprecedented size and buying power." Neil McKendrick, John Brewer, and John Harold Plumb, *The Birth of a Consumer Society: The Commercialization of Eighteenth-Century England* (Bloomington: Indiana University Press, 1982), 5.

36 **The first full-color fashion image:**, McKendrick et al., 46.

38 **"It is often said":** Raymond Williams, "Advertising: The Magic System," *The Advertising and Consumer Culture Reader*, Matthew P. McAllister and Joseph Turow, eds. (New York: Routledge, 2009), 13–24.

39 **"The present rage":** McKendrick et al., 53.

39 **"Is fashion in fact":** Fernand Braudel, *Civilization and Capitalism, 15th–18th Century: The Structure of Everyday Life* (Berkeley: University of California Press, 1979), 323.

41 **"What's necessary":** Quoted in Elaine Showalter's introduction to Émile Zola, *Au Bonheur des Dames (The Ladies' Delight)*, trans. Robin Buss (New York: Penguin Classics, 2007), 415.

42 **"Dazzling and sensuous, the Bon Marché":** Michael Miller, *The Bon Marché: Bourgeois Culture and the Department Store, 1869–1920* (Princeton, NJ: Princeton University Press, 1981), 162.

47 **"department store thefts":** Quoted in Miller, 202–8.

50 **"a pitcher plant":** Quoted in M. Jeffrey Hardwick, *Mall Maker: Victor Gruen, Architect of an American Dream* (Philadelphia: University of Pennsylvania Press, 2004), 33.

51 **"avenues of horror":** Quoted in Malcolm Gladwell, "The Terrazzo Jungle. Fifty Years Ago, the Mall Was Born. America Would Never Be the Same," *The New Yorker* 15 (2004).

54 **"Southdale was not a suburban alternative":** Ibid.

54 **"The service done by the Fort Worth":** Quoted in Hardwick, 181.

55 **"giant shopping machine":** Quoted in Hardwick, 211.

Chapter 2. Music

71 **"We enjoy strawberry cheesecake":** Steven Pinker, *How the Mind Works* (New York: Norton, 1999), 535.

71 **"The presence of music":** Nicholas J. Conard, Maria Malina, and Susanne C. Münzel, "New Flutes Document the Earliest Musical Tradition in Southwestern Germany," *Nature* 460:7256 (2009), 739.

71 **Others take the sexual conquests:** A fine overview of the arguments for the evolutionary roots of music can be found in Daniel J. Levitin's *This Is Your Brain on Music: Understanding a Human Obsession* (London: Atlantic Books Ltd., 2011).

74 **"We wish to explain," the brothers:** Imad Samir, *Allah's Automata: Artifacts of the Arab-Islamic Renaissance (800-1200)* (Berlin: Hatje Cantz, 2015), 68–86.

80 **"Using the Jacquard loom":** James Essinger, *Jacquard's Web: How a Hand-Loom Led to the Birth of the Information Age* (New York: Oxford University Press, Kindle edition), 38.

82 **"You are aware":** Essinger, 47.

82 **When his collaborator Ada Lovelace:** Quoted in Johnson, *How We Got to Now: Six Innovations that Made the Modern World*, 249.

85 **A roster of instruments:** Tim Carter, "A Florentine Wedding of 1608," *Acta Musicologica* 55, Fasc. 1 (1983), 95.

87 **The chips might have followed:** It would seem that typewriter-style keyboards are a condition of possibility for advanced computers, almost the way capturing and transmitting electricity was a condition of possibility for the lightbulb. The former needs to come before the latter. And yet, strangely enough, computers were invented before typewriters, if you consider Babbage's analytic engine to be the first computer. Babbage figured out how to swap algorithms in and out of random access memory before the rest of us figured out how to strike a few keys with our fingers and make letters appear on a page.

88 **"From a mechanical point of view":** Michael H. Adler, *The Writing Machine* (London: Allen and Unwin, 1973), 5.

91 **The very first long-distance:** For more on the talking drums, see James Gleick, *The Information: A History, a Theory, a Flood* (New York: Vintage, 2012).

97 **"The Ballet began":** Richard Rhodes, *Hedy's Folly: The Life and Breakthrough Inventions of Hedy Lamarr, the Most Beautiful Woman in the World* (New York: Doubleday, 2011), 68.

97 **"outsacked the Sacre":** Paul Lehrman, "Blast from the Past," *Wired*, November 1, 1999, http://www.wired.com/1999/11/ballet.

97 **"From this moment on":** Ibid.

101 **Antheil later wondered:** Quoted in Anna Corey, "How 'The Bad Boy of Music' and 'The Most Beautiful Girl in the World' Catalyzed a Wireless Revolution—in 1941." http://people.seas.harvard.edu/~jones/cscie129/nu_lectures/lecture7/hedy/lemarr.htm.

101 **"We shall see orchestral machines":** Quoted in Lehrman, "Blast from the Past."

103 **"I was allowed to sing":** All quotations from Oram cited in Jo Hutton, "Daphne Oram: Innovator, Writer and Composer," *Organised Sound* 8, 49–56.

Chapter 3. Taste

111 **Archeologists in Syria:** Daniel T. Potts, *Mesopotamian Civilization: The Material Foundations* (Ithaca, NY: Cornell University Press, 1997), 269.

111 **Somehow, in an era before compasses:** "By the turn of the millennium, they crop up in the records of cities spread around its shores: Marseilles, Barcelona, Ragusa. Some spices arrived via Byzantium and the Black Sea, following the Danube to eastern and central Europe, but the greatest volume of traffic passed through Alexandria and the Levant to Italy. From Italy a number of routes led north over the Alpine passes toward France and Germany." Jack Turner, *Spice: The History of a Temptation* (New York: Knopf, Kindle edition), Kindle location 1108.

113 **"King of England, Scotland":** Ibid., loc. 956.

115 **"Nowhere is the history of East and West":** Ibid., loc. 5879.

116 **But the real treasure:** Filipe Castro, "The Pepper Wreck, an Early 17th-Century Portuguese Indiaman at the Mouth of the Tagus River, Portugal," *International Journal of Nautical Archaeology* 32:1 (August 2003), 6–23.

116 **The custom of paying:** "As late as 1937, the king of England received rent from the mayor of Launceston consisting of a hundred shillings and one pound of pepper— the mayor might have reflected that this particular rent ceiling had proved very much to his financial advantage. When Prince Charles crossed the River Tamar in 1973 to take symbolic possession of the Duchy of Cornwall, his tribute included a pound of pepper. According to the *OED*, a token pepper rent remained a form of payment until the end of the nineteenth century." Turner, Kindle location 1932.

116 **When the Portuguese queen Isabella:** Marjorie Shaffer, *Pepper: A History of the World's Most Influential Spice* (New York: St. Martin's Griffin, 2013), Kindle location 439.

118 **"a variety of spiced desserts":** Turner, Kindle location 1626.

118 **"There is no year in which India":** Quoted in Abraham Eraly, *The First Spring: The Golden Age of India* (New Delhi: Penguin Books India, 2011), 244.

118 **The Dutch East India Company:** The Dutch East India Company also served the aims of the Dutch state itself, and not simply the company's investors. "The VOC had a dual purpose, in order to obtain commercial as well as political objectives: to trade in Asia, but at the same time to make war on the enemies of the newborn Dutch Republic, Spain and Portugal. It received sovereign rights within its chartered area, such as the right to declare war, enter into diplomatic negotiations or sign treaties with local rulers, or to organize and execute military operations. As an institution, it was virtually a state within the state." Vincent C. Loth, "Armed Incidents and Unpaid Bills: Anglo-Dutch Rivalry in the Banda Islands in the Seventeenth Century," *Modern Asian Studies* 29:4 (1995), 708.

119 **"The Mallayans":** Shaffer, Kindle location 2201.

119 **"Nothing is more certain":** Ibid., loc. 2228.

119 **In what the historian Vincent Loth:** Loth, 725.

122 **One theory holds:** Tim Ecott, *Vanilla: Travels in Search of the Ice Cream Orchid* (New York: Grove/Atlantic, Inc., 2004, Kindle edition), 75–76.
124 **"I knew from experience":** Lewis A. Maverick, "Pierre Poivre: Eighteenth Century Explorer of Southeast Asia," *Pacific Historical Review* 10:2 (1941), 171.
125 **"Nearly two hundred years later":** Turner, Kindle locations 5718–5719.
125 **"Perhaps the earliest known use":** Ecott, 6.
126 **"Everyone uses this confection":** Quoted in Ecott, 21–22.
126 **"[My butler] informs me":** Ibid., 83.
129 **Vanilla plants:** Joseph Arditti, A. N. Rao, and H. Nair, "Hand-Pollination of Vanilla: How Many Discoverers," *Orchid Biology: Reviews and Perspectives* 11 (2009), 233–49.
129 **"This clever boy had realized":** Quoted in Ecott, 151.
130 **"The very man":** Ecott, 106.
131 **In medieval England:** "When Edward I returned to London from the wars in Wales at the end of the thirteenth century, his officers spent more than £1,775 on spices out of a total expenditure on luxuries of just under £10,000—a staggering sum, even taking into account that many of his 'spices' included items such as oranges and sugar. To put the figure into perspective, his spice expenditure was about the same as the total annual income of an earl." Turner, Kindle location 2731.
131 **"To limit their function":** Wolfgang Schivelbusch, *Tastes of Paradise: A Social History of Spices, Stimulants, and Intoxicants* (New York: Pantheon Books, 1992), 6.
132 **According to Turner:** Turner, Kindle location 2568.
133 **"a decoction of vanilla beans":** Ecott, 16–17.
134 **"electuary that I made":** Paul Delany, "Constantinus Africanus' 'De Coitu': A Translation," *Chaucer Review* (1969), 55–65.
134 **"No fewer than 342 impotent men":** Ecott, 23.
136 **"Doubtless a vigorous error":** George Eliot, *Middlemarch* (New York: Penguin Classics, 2002), 478.
136 **"a powder of cinnamon":** Turner, Kindle location 3527.
137 **"three galleys put in at Genoa":** Quoted in Tom Standage, *An Edible History of Humanity* (New York: Bloomsbury, 2009), 82.
139 **"good things of his creation":** Shaffer, Kindle location 1112.
142 **"It is remarkable that [pepper's]":** Cited in Standage, 71.

Chapter 4. Illusion

147 **"a hoarse and terrible tone":** Laurent Mannoni and Ben Brewster, "The Phantasmagoria," *Film History* 8:4 (1996), 392.
149 **"something you have never seen before":** Deac Rossell, *The 19th Century German Origins of the Phantasmagoria Show* (February 16, 2001), 3.
149 **"In dying," Deac Rossell writes:** Ibid, 4.
149 **"As soon as the peroration":** Ibid, 5.
151 **"This is the night . . . phantasmagoria":** Stefan Andriopoulos, *Ghostly Apparitions: German Idealism, the Gothic Novel, and Optical Media* (New York: Zone Books, 2013, Kindle edition), Kindle locations 19–21.
153 **"We sit in a boundless Phantasmagoria":** Quoted in Kevin Hetherington, *Cap-*

italism's Eye: Cultural Spaces of the Commodity (New York: Routledge, 2011), 244.

153 **"As against this, the commodity-form":** Karl Marx, *Capital: Volume One* (New York: Vintage Books, 1977), 165.

155 **"[The] phenomena were produced":** David Brewster, *Letters on Natural Magic* (London: Chatto & Windus, 1883), 85.

157 **"At all times, curiosity":** Richard Daniel Altick, *The Shows of London* (Cambridge, MA: Harvard University Press, 1978), 3.

157 **"The eye," he wrote:** Brewster, 21.

161 **"an entire new Contrivance":** Altick, 131.

162 **"The greatest IMPROVEMENT":** Erkki Huhtamo, *Illusions in Motion: Media Archaeology of the Moving Panorama and Related Spectacles* (Cambridge, MA: MIT Press, 2013), 1.

162 **"No device . . . has approached":** Charles Taylor, *The Literary Panorama* (London: Cox, Son, and Baylis, 1810), 447.

163 **"It is a delightful characteristic":** Charles Dickens, *Household Words*, Vol. 1, 1850, 73–77.

164 **"musicians, snipers, cavalry":** John F. Ohl and Joseph Earl Arrington, "John Maelzel, Master Showman of Automata and Panoramas," *Pennsylvania Magazine of History and Biography* 84:1 (1960), 79.

165 **"The city was before us":** Ibid.

165 **"A musket machine":** Ibid.

167 **"Novel Mechanical and Pictorial Exhibition":** Altick, 131.

167 **"TRADITIONARY GHOST WORK!":** Ibid.

169 **"You are allowed to look":** Altick, 231.

178 **"no one thought of clothing":** Neal Gabler, *Walt Disney* (New York: Knopf Doubleday, 2006, Kindle edition), Kindle locations 3758–3763.

178 **pencil tests:** "In short order, Walt installed a Moviola in a cramped, stifling, windowless closet that was soon dubbed the 'sweatbox.' Hunched over the tiny screen, no more than four inches by four inches, Walt and the animator would view and analyze the action by the hour, over and over and over again, trying to determine what would make it right, make it funnier. 'I think it is astounding that we were the first group of animators, so far as I can learn, who ever had the chance to study their own work and correct its errors before it reached the screen.'" Gabler, Kindle locations 3644–3654.

179 **"[The device] consists of four":** Andrew R. Boone, "The making of *Snow White and the Seven Dwarfs, Popular Science*, January 1938, 50, http://blog.modernmechanix.com/the-making-of-snow-white-and-the-seven-dwarfs/.

181 **"No animated cartoon":** Gabler, Kindle locations 5823–5828.

181 **"Before the film was shown":** "They were the scenes in which the audience would be invited to cry along with the dwarfs, an emotional province that animations had not previously entered, and they would constitute the major test of the film's effectiveness, though by this time Walt had little doubt they would succeed. 'There is going to be a lot of sympathy for these little fellows,' he said at a story conference that July. 'We can tear their [the audience's] hearts out if we want to by putting in a little crying.' Frank Thomas, Fred Moore's onetime assistant, was given the as-

signment of animating the dwarfs' grief from Albert Hurter's drawings, and he animated it with as little movement as possible—basically held poses with tears crawling down the dwarfs' cheeks and, as Walt had instructed at a story meeting, 'concentrating on Grumpy when he breaks down and starts to cry,' cracking his stoic facade." Gabler, Kindle locations 5714–5723.

183 **"celebrity was . . . acknowledgement of achievement":** Fred Inglis, *A Short History of Celebrity* (Princeton, NJ: Princeton University Press, 2010, Kindle edition), 53.

183 **"We still try to make our celebrities":** Daniel J. Boorstin, *The Image: A Guide to Pseudo-Events in America* (New York: Harper & Row, 1964), 74.

184 **"knowability combined with distance":** "This is the powerful contradiction at the heart of our phenomenon. It combines knowability with distance. Political leader and cinema star are intensely familiar (one of the family) by way of the cinema screen, and (at first) by way of their voices on the living room radio, but physically and in terms of how we all need to feel the directness of experience, they have the remoteness of the supernatural." Inglis, 11.

Chapter 5. Games

187 **Over the years, his name:** William Caxton, *Game and Playe of the Chesse* (Charleston, SC: BiblioBazaar, LLC, 2007), Google Scholar, xxiv.

187 **"The original was not just copied":** Jacobus de Cessolis, *The Book of Chess* (New York: Italica Press, 2008, Kindle edition), Kindle locations 136–143.

189 **"Both knights have three":** de Cessolis, Kindle locations 113–114.

189 **"The king has dominion":** de Cessolis, Kindle locations 1245–1251.

189 **"If women want to remain chaste":** de Cessolis, Kindle locations 1318–1319.

190 **"The chess allegory":** Jenny Adams, *Power Play: The Literature and Politics of Chess in the Late Middle Ages,* The Middle Ages Series (Philadelphia: University of Pennsylvania Press, 2006, Kindle edition), Kindle locations 420–425.

190 **Authority no longer trickled:** "When someone in the game says 'Check!' to a king, he's signifying a demand for rights. It is the same as saying, 'Your honor, allow me to pass.' The king has to grant the move unless, through wisdom or power, he can defend himself.

"It is often the case that when a knight, servant, nobleman, or commoner feels that an injustice has been done to him, or an excessive constraint put on him, if he can't otherwise get revenge on the king, he waits until the king comes with him to the battlefield, and then he flees, leaving the king exposed to the enemy." De Cessolis, Kindle locations 1285–1289.

192 **"Kings cajoled":** David Shenk, *The Immortal Game: A History of Chess* (New York: Knopf Doubleday, 2006, Kindle edition), Kindle locations 221–225.

193 **"Can [a] machine play chess":** Cited in Andrew Hodges, *Alan Turing: The Enigma* (Princeton, NJ: Princeton University Press, 2012), 336.

193 **Some cognitive scientists compared:** "Just as biologists need model organisms to explore genetics," writes the University of Waterloo's Neil Charness, "so too do cognitive scientists need model task environments to study adaptive cognitive mechanisms. Chess playing provides a rich task environment that taps many cogni-

tive processes, ranging from perception, to memory, to problem solving." Quoted in Shenk, Kindle location 1742.

195 **"[The] board had an Intemperance":** Mary Pilon, *The Monopolists: Obsession, Fury, and the Scandal Behind the World's Favorite Board Game* (New York: Bloomsbury Publishing, 2015, Kindle edition), Kindle locations 729–735.

196 **"Representative money":** Lizzie J. Magie, *Land and Freedom: An International Record of Single Tax Progress*, Vol. II (1904).

197 **"The little landlords take":** Ibid.

199 **A lesson in the abuses:** Magie would finally warrant a grudging acknowledgment on starting with the 1991 edition of Monopoly: "It all started back in 1933 when Charles B. Darrow of Germantown, Pennsylvania was inspired by The Landlord's Game to create a new diversion to entertain himself while he was unemployed." Pilon, Kindle locations 3501–3510.

201 **"Like the Bible":** Shenk, Kindle locations 329–335.

202 **The practice of announcing the king's:** de Cessolis, Kindle locations 50–55.

203 **"The wearingness which players experienced":** Quoted in Shenk, Kindle location 874.

205 **Enraged at this deception:** Artur Ekert, "Complex and Unpredictable Cardano," *International Journal of Theoretical Physics* 47:8 (2008), 2102.

205 **"hot tempered, single-minded":** Ibid.

208 **"The pundits":** Keith Devlin, *The Unfinished Game: Pascal, Fermat, and the Seventeenth-Century Letter That Made the World Modern* (New York: Basic Books, 2008, Kindle edition), Kindle location 99.

208 **Before Cardano:** "The first known attempt to discern numerical patterns in games of chance seems to have come around 960, when Bishop Wibold of Cambrai correctly enumerated the 56 outcomes that can arise when three dice are thrown simultaneously: 1, 1, 1; 1, 1, 2; 2, 3, 5; and so on. A thirteenth-century Latin poem, De Vetula, listed the 216 (= 6 × 6 × 6) outcomes that may result when three dice are thrown in succession." Devlin, Kindle location 141.

209 **"[Each] had two rounded sides":** Devlin, Kindle location 132.

209 **Two sets of statutes:** Steven A. Epstein, *Wage Labor and Guilds in Medieval Europe* (Chapel Hill: University of North Carolina Press 1991), 127–28.

210 **"a stuffed ball":** John Fox, *The Ball: Discovering the Object of the Game* (New York: Harper Perennial, 2012), Kindle location 355.

211 **"Jumping and bouncing":** Quoted in Fox, Kindle location 1269.

213 **"I don't understand":** John Tully, *The Devil's Milk: A Social History of Rubber* (New York University Press, 2011, Kindle edition), 31.

214 **"The demand for rubber . . . in Peru":** Tully, 23.

214 **But the Mesoamericans:** "In the 1940s, Paul Stanley, a botanist at the Field Museum of Natural History in Chicago, identified the vine as *Ipomoea alba*, a kind of night-blooming morning glory commonly known as moon vine or moonflower. Recent studies show that when latex from Castilla elastica is boiled with the juice of moon vine, sulfonic acids that occur naturally in the vine increase the plasticity and elasticity of the rubber and produce a degree of vulcanization." Fox, Kindle location 1300.

216 **"It should *demonstrate*":** J. M. Graetz, "The Origin of Spacewar!," *Creative Computing*, August 1981, www.wheels.org/spacewar/creative/SpacewarOrigin.html.

217 **"The game of Spacewar!":** Stewart Brand, "Spacewar!," *Rolling Stone*, December 7, 1972, www.wheels.org/spacewar/stone/rolling_stone.html.

217 **"Using data from the *American Ephemeris*":** Graetz, "The Origin of Spacewar!"

222 **"mechanically well made":** Edward O. Thorp, "Wearable Computers," *Digest of Papers*, Second International Symposium on. 1998.

223 **"It had perhaps a hundred thousand":** Ibid.

225 **"As we worked and during":** Ibid.

229 **"The computer's techniques":** Ken Jennings, "My Puny Human Brain," *Slate*, *Newsweek* Interactive Co. LLC, February 2012.

Chapter 6. Public Space

233 **"Someone might play":** Leopold S. Launitz-Schurer, "Slave Resistance in Colonial New York: An Interpretation of Daniel Horsmanden's New York Conspiracy," *Phylon* (1960), 144.

234 **"young gentlemen inclined":** Edwin G. Burrows and Mike Wallace, *Gotham: A History of New York City to 1898* (New York: Oxford University Press, 1998), 165.

235 **"more than happy":** Christine Sismondo, *America Walks into a Bar: A Spirited History of Taverns and Saloons, Speakeasies and Grog Shops* (Oxford University Press, 2011, Kindle edition), Kindle locations 751 752.

236 **"a penny dram of a penny worth":** Launitz-Schurer, 148.

236 **"For people who have been":** Launitz-Schurer, 151.

239 **"For one [coin] you can drink":** Iain Gately, *Drink: A Cultural History of Alcohol* (New York: Penguin Publishing Group, 2008, Kindle edition), 35.

239 **"the stages of travel":** W. C. Firebaugh, *The Inns of Greece and Rome* (Chicago: Pascal Covici, 1928), 122.

240 **The census of 1577:** Gately, 110–111.

240 **"[Pubs] were run":** Gately, 85–86.

241 **"becoming in many places the nurseries":** David W. Conroy, *In Public Houses: Drink and the Revolution of Authority in Colonial Massachusetts* (Chapel Hill: University of North Carolina Press, 1995), 204.

242 **"beautiful young men . . . with bright eyes":** Sismondo, Kindle location 4402.

243 **"kissing other men":** Lillian Faderman and Stuart Timmons, *Gay L.A.: A History of Sexual Outlaws, Power Politics, and Lipstick Lesbians* (New York: Basic Books, 2006, Kindle edition), Kindle locations 1864–1872.

245 **"paralyzing political skepticism":** Craig Calhoun, *Contemporary Sociological Theory* (New York: John Wiley & Sons, 2012), 256.

245 **Habermas observed:** Jürgen Habermas, *The Structural Transformation of the Public Sphere: An Inquiry into a Category of Bourgeois Society* (Cambridge, MA: MIT Press, 1989), 26.

246 **"a kind of social intercourse":** Ibid.

247 **"When insects feed on caffeine-spiked nectar":** Carl Zimmer, "How Caffeine Evolved to Help Plants Survive and Help People Wake Up," *The New York Times* (September 4, 2014).

248 **"syrup of soot":** Matthew Green, "The Lost World of the London Coffeehouse," *Public Domain Review* 7 (2013).

251 **"Where it is most apparent":** Quoted in J. H. Brindley, "Commercial Aspects of Coffee," in *Coffee and Tea Industries and the Flavor Field*, 37.

252 **Amsterdam, at that point:** Green, 2013.

252 **"At the Bedford Coffeehouse":** Ibid.

254 **"All accounts of gallantry, pleasure":** Quoted in Brindley, "Commercial Aspects of Coffee."

254 **"Unexpectedly wide-ranging discussions":** Green, 2013.

255 **"A Swedish owl":** Richard Daniel Altick, *The Shows of London* (Cambridge, MA: Harvard University Press, 1978), 15.

257 **"not only for the inspection":** Ibid, 26.

258 **"It reason seems that liberty":** Bonnie Calhoun, "Shaping the Public Sphere: English Coffeehouses and French Salons and the Age of the Enlightenment," *Colgate Academic Review* 3:1 (2012), 83.

258 **"First, Gentry, Tradesmen":** Ibid.

258 **"Blue and Green Ribbons":** John Macky, "A Journey Through England. In Familiar Letters. From a Gentleman Here, to his Friends Abroad (1714)," *Eighteenth-Century Coffee-House Culture*, Vol. I, Ellis Markman, ed. (London: Pickering & Chatto, 2006), 339.

260 **In 1620, when the first Pilgrims:** William Bradford *Of Plymouth Plantation* (Chelmsford, MA: Courier Corporation, 2012), 62.

264 **"I could hardly believe my eyes":** Quoted in Cian Duffy and Peter Howell, *Cultures of the Sublime: Selected Readings, 1750–1830* (New York: Palgrave Macmillan, 2011).

267 **The wellsprings that fed this new form:** Wordsworth had seen a prophesy of this future back in the late 1700s, during a visit to the St. Bartholomew fair, famously described in a passage from *The Prelude* that includes a sly reference to Merlin's Mechanical Museum:

All moveables of wonder, from all parts,
Are here—Albinos, painted Indians, Dwarfs,
The Horse of knowledge, and the learned Pig,
The Stone-eater, the man that swallows fire,
Giants, Ventriloquists, the Invisible Girl,
The Bust that speaks and moves its goggling eyes,
The Wax-work, Clock-work, all the marvellous craft
Of modern Merlins, Wild Beasts, Puppet-shows,
All out-o'-the-way, far-fetched, perverted things,
All freaks of nature, all Promethean thoughts
Of man, his dullness, madness, and their feats
All jumbled up together, to compose
A Parliament of Monsters.

268 **"In his mind's eye":** Also from the same handbook: "Not only will there be satisfaction of the natural and rational curiosity of an observer, in the sight of creatures

strange to our clime and notions, brought from different lands, of which from our childhood we all have heard so much, but his imagination will be gratified by being carried out to them by their denizens being now actually before him." Quoted in Robert W. Jones, "The Sight of Creatures Strange to Our Clime: London Zoo and the Consumption of the Exotic," *Journal of Victorian Culture* 2:1 (1997), 6.

268 **she wrote in her diary:** Randal Keynes, *Darwin, His Daughter, and Human Evolution* (New York: Penguin, 2002, Kindle edition), Kindle locations 784–786.

269 **"She threw herself on her back":** Ibid.

269 **"Let man visit Ourang-outang":** Ibid., Kindle locations 854–858.

271 **"An enormous task lay":** Eric Ames, *Carl Hagenbeck's Empire of Entertainments* (Seattle: University of Washington Press, 2008), 163.

272 **At the summit of the "Northern Plateau":** Ibid., 180.

272 **"When you approach the Lions' Ravine":** Ibid., 184.

273 **"Disneyland is presented":** Jean Baudrillard, *Simulacra and Simulation* (Ann Arbor, MI: University of Michigan Press, 1994), 12.

Conclusion

280 **The model has several variants:** R. A. Rescorla and A. R. Wagner, "A Theory of Pavlovian Conditioning: Variations in the Effectiveness of Reinforcement and Nonreinforcement," in A. H. Black and W. F. Prokasy, eds., *Classical Conditioning II* (New York: Appleton-Century-Crafts, 1972), 64–99.

281 **A new theory proposes that dopamine:** Andrew Barto, Marco Mirolli, and Gianluca Baldassarre, "Novelty or Surprise?," *Frontiers in Psychology* 4 (2013).

281 **The computer scientist Jürgen Schmidhuber:** Rafal Salustowicz and Jürgen Schmidhuber, "Probabilistic Incremental Program Evolution," *Evolutionary Computation* 5:2 (1997), 123–41.

283 ***because* of our instincts and nature:** I suspect this strange relationship between genetic determination and cultural exploration that play activates may be one reason that Johan Huizinga divorced his otherwise brilliant 1938 analysis of play, *Homo Ludens*, from any biological grounding. "The aim of the present full-length study," he wrote, "is to try to integrate the concept of play into that of culture. Consequently, play is to be understood here not as a biological phenomenon but as a cultural phenomenon." The surprise instinct suggests that the two domains are in fact profoundly intertwined, perhaps more so than in any other form of human experience. Huizinga, Johan. *Homo Ludens: A Study of the Play-Element in Culture* (Boston: Beacon Press, Kindle edition), Kindle locations 48–49.

BIBLIOGRAPHY

Adburgham, Alison. *Shopping in Style: London from the Restoration to Edwardian Elegance.* New York: Thames and Hudson, 1979.

Adler, Michael H. *The Writing Machine.* London: Allen and Unwin, 1973.

Altick, Richard Daniel. *The Shows of London.* Cambridge, MA: Harvard University Press, 1978.

Ames, Eric. *Carl Hagenbeck's Empire of Entertainments.* Seattle: University of Washington Press, 2008.

Anbinder, Tyler. *Five Points: The 19th-Century New York City Neighborhood that Invented Tap Dance, Stole Elections, and Became the World's Most Notorious Slum.* New York: Simon & Schuster, 2001.

Antheil, George, and Marthanne Verbit. *Bad Boy of Music.* New York: S. French, 1990.

Arditti, Joseph, A. N. Rao, and H. Nair. "Hand-pollination of Vanilla: How Many Discoverers." *Orchid Biology: Reviews and Perspectives* XI (New York Botanical Garden, 2009): 233–49.

Babbage, Charles. *Passages from the Life of a Philosopher.* Cambridge University Press, 2011.

Barbon, Nicholas. *A Discourse of Trade.* Printed by Tho. Milbourn for the author, London, 1905.

Barto, Andrew, Marco Mirolli, and Gianluca Baldassarre. "Novelty or Surprise?" *Frontiers in Psychology* 4 (2013).

Baudrillard, Jean. *Simulacra and Simulation.* Ann Arbor, MI: University of Michigan Press, 1994.

Behbehani, Abbas M. "The Smallpox Story: Life and Death of an Old Disease." *Microbiological Reviews* 47:4 (1983): 455.

Benson, John, and Laura Ugolini. *A Nation of Shopkeepers: Five Centuries of British Retailing.* London: IB Tauris, 2003.

Benvenuto, Sergio. "Fashion: Georg Simmel." *Journal of Artificial Societies and Social Simulation* 3:2 (2000).

Billing, Jennifer, and Paul W. Sherman. "Antimicrobial Functions of Spices: Why Some Like It Hot." *Quarterly Review of Biology* (1998): 3–49.

Blumer, Herbert. "Fashion: From Class Differentiation to Collective Selection." *Sociological Quarterly* (1969): 275–91.

Boorstin, Daniel J. *The Image: A Guide to Pseudo-events in America.* New York: Harper & Row, 1964.

Bowlby, Rachel. *Carried Away: The Invention of Modern Shopping.* New York: Columbia University Press, 2001.

Bradford, William. *Of Plymouth Plantation.* Chelmsford, MA: Courier Corporation, 2012.

Braudel, Fernand. *Civilization and Capitalism, 15th–18th Century: The Structure of Everyday Life.* Berkeley: University of California Press, 1979.

Brewster, David. *Letters on Natural Magic.* London: Chatto & Windus, 1883.

Bryson, Bill. *At Home: A Short History of Private Life.* New York: Random House, 2010.

Burrows, Edwin G., and Mike Wallace. *Gotham: A History of New York City to 1898.* New York: Oxford University Press, 1998.

Calhoun, Bonnie. "Shaping the Public Sphere: English Coffeehouses and French Salons and the Age of the Enlightenment." *Colgate Academic Review* 3:1 (2012): 7.

Calhoun, Craig. *Contemporary Sociological Theory.* New York: John Wiley & Sons, 2012.

Carter, Tim. "A Florentine Wedding of 1608." *Acta Musicologica* 55. Fasc. 1 (1983): 89–107.

Castro, F. "The Pepper Wreck, An Early 17th-Century Portuguese Indiaman at the Mouth of the Tagus River, Portugal." *International Journal of Nautical Archaeology* 32:1 (August 2003): 6–23.

Caxton, William. *Game and Playe of the Chesse.* Charleston, SC: BiblioBazaar, LLC, 2007.

Collier, Bruce, and James MacLachlan. *Charles Babbage: And the Engines of Perfection.* Oxford University Press, 2000.

Conard, Nicholas J., Maria Malina, and Susanne C. Münzel. "New Flutes Document the Earliest Musical Tradition in Southwestern Germany." *Nature* 460.7256 (2009): 737–40.

Conroy, David W. *In Public Houses: Drink & the Revolution of Authority in Colonial Massachusetts.* Chapel Hill: University of North Carolina Press, 1995.

Corey, Anna. "How 'The Bad Boy of Music' and 'The Most Beautiful Girl in the World' Catalyzed a Wireless Revolution—in 1941." 1997, http://people.seas.harvard.edu/~jones/cscie129/nu_lectures/lecture7/hedy/lemarr.htm.

Davis, William Stearns. *Readings in Ancient History: Illustrative Extracts from the Sources.* Vol. 1. Boston: Allyn and Bacon, 1913.

Defoe, Daniel. *The Complete English Tradesman.* Vol. 17. DA Talboys, 1841.

Delany, Paul. "Constantinus Africanus' 'De Coitu': A Translation." *Chaucer Review* (1969): 55–65.

Devlin, Keith. *The Unfinished Game: Pascal, Fermat, and the Seventeenth-Century Letter that Made the World Modern*. New York: Basic Books, 2010.

Diamond, Jared. *Guns, Germs, and Steel: The Fates of Human Societies*. New York: W. W. Norton & Company, 1999.

Douthwaite, Julia V., and Daniel Richter. "The Frankenstein of the French Revolution: Nogaret's Automaton Tale of 1790." *European Romantic Review* 20:3 (July 2009): 381–411.

Duffy, Cian, and Peter Howell. *Cultures of the Sublime: Selected Readings, 1750–1830*. New York: Palgrave Macmillan, 2011.

Dumper, Michael, and Bruce E. Stanley. *Cities of the Middle East and North Africa: A Historical Encyclopedia*. Santa Barbara, CA: ABC-CLIO, 2007.

Ecott, Tim. *Vanilla: Travels in Search of the Ice Cream Orchid*. New York: Grove Press, 2005.

Ekert, Artur. "Complex and Unpredictable Cardano." *International Journal of Theoretical Physics* 47:8 (2008): 2101–19.

Eliot, George. *Middlemarch*. New York: Penguin Classics, 2002.

Epstein, Steven A. *Wage Labor and Guilds in Medieval Europe*. Chapel Hill: University of North Carolina Press, 1991.

Eraly, Abraham. *The First Spring: The Golden Age of India*. New Delhi: Penguin Books India, 2011.

Essinger, James. *Jacquard's Web: How a Hand-Loom Led to the Birth of the Information Age*. New York: Oxford University Press, 2004.

Faderman, Lillian, and Stuart Timmons. *Gay L.A.: A History of Sexual Outlaws, Power Politics, and Lipstick Lesbians*. New York: Basic Books, 2006.

Farmer, Henry George. *The Organ of the Ancients: From Eastern Sources (Hebrew, Syriac and Arabic)*. W. Reeves, 1931.

Firebaugh, W. C. *The Inns of Greece and Rome*. Chicago: Pascal Covici, 1928.

Fowler, Charles B. "The Museum of Music: A History of Mechanical Instruments." *Music Educators Journal* 54:2 (October 1967): 45–49.

Fox, John. *The Ball: Discovering the Object of the Game*. New York: Harper Perennial, 2012.

Freedman, Paul. *Out of the East: Spices and the Medieval Imagination*. New Haven, CT: Yale University Press, 2008.

Gabler, N. "Toward a New Definition of Celebrity" (2004). Retrieved April 11, 2008. http://learcenter.org/pdf/Gabler.pdf.

Gabler, Neal. *Walt Disney: The Triumph of the American Imagination*. New York: Knopf Doubleday, 2006.

Gladwell, Malcolm. "The Terrazzo Jungle. Fifty Years Ago, the Mall Was Born. America Would Never Be the Same." *The New Yorker* (March 15, 2004).

Gleick, James. *The Information: A History, A Theory, A Flood*. New York: Vintage, 2012.

Gordon, Margaret Maria. *The Home Life of Sir David Brewster*. Edinburgh: D. Douglas, 1881.

Green, Matthew. "The Lost World of the London Coffeehouse." *Public Domain Review* 7 (2013).

Gruen, Victor. *The Heart of Our Cities: The Urban Crisis: Diagnosis and Cure*. London: Thames and Hudson, 1965.

Habermas, Jürgen, translated by Thomas Burger. *The Structural Transformation of the Public Sphere*. Cambridge, MA: MIT Press (1989): 85–92.

Hardwick, M. Jeffrey. *Mall Maker: Victor Gruen, Architect of an American Dream*. Philadelphia: University of Pennsylvania Press, 2004.

Hetherington, Kevin. *Capitalism's Eye: Cultural Spaces of the Commodity*. New York: Routledge, 2011.

Hill, Donald Routledge. "Mechanical Engineering in the Medieval Near East." *Scientific American* 264:5 (1991): 100–105.

Hodges, Andrew. *Alan Turing: The Enigma*. Princeton, NJ: Princeton University Press, 2012.

Huebner, Andrew J. "The Conditional Optimist: Walt Disney's Postwar Futurism." *The Sixties: A Journal of History, Politics and Culture* 2:2 (2009): 227–44.

Huhtamo, Erkki. *Illusions in Motion: Media Archaeology of the Moving Panorama and Related Spectacles*. Cambridge, MA: MIT Press, 2013.

Huizinga, Johan. *Homo Ludens*. New York: Routledge, 2014.

Hutton, Jo. "Daphne Oram: Innovator, Writer and Composer." *Organised Sound* 8:01 (2003): 49–56.

Jacobus de Cessolis, translated and edited by H. L. Williams. *The Book of Chess*. New York: Italica Press, 2008.

Johnson, Steven. *How We Got to Now: Six Innovations that Made the Modern World*. New York: Penguin, 2014.

Jones, Christopher. "The Rubber Ball Game: A Universal Mesoamerican Sport." *Expedition: The Magazine of the University of Pennsylvania* 27:2 (1985): 44–52.

Jones, Robert W. "The Sight of Creatures Strange to Our Clime: London Zoo and the Consumption of the Exotic." *Journal of Victorian Culture* 2:1 (1997): 1–26.

Keynes, Randal. *Darwin, His Daughter, and Human Evolution*. New York: Penguin, 2002.

Launitz-Schurer, Leopold S. "Slave Resistance in Colonial New York: An Interpretation of Daniel Horsmanden's New York Conspiracy." *Phylon* (1980): 137–52.

Lehrman, Paul. "Blast from the Past." *Wired* (November 1, 1999). http://www.wired.com/1999/11/ballet.

Levitin, Daniel J. *This Is Your Brain on Music: Understanding a Human Obsession*. New York: Atlantic Books Ltd, 2011.

Lombard, M., P. P. Pastoret, and A. M. Moulin. "A Brief History of Vaccines and Vaccination." *Revue Scientifique et Technique-Office International des Epizooties* 26:1 (2007): 29.

Loth, Vincent C. "Armed Incidents and Unpaid Bills: Anglo-Dutch Rivalry in the Banda Islands in the Seventeenth Century." *Modern Asian Studies* 29:04 (1995): 705–40.

Lyons, Jonathan. *The House of Wisdom: How the Arabs Transformed Western Civilization*. New York: Bloomsbury Publishing, 2011.

Manning, Peter. "The Oramics Machine: From Vision to Reality." *Organised Sound* 17:02 (2012): 137–47.

Mannoni, Laurent, and Ben Brewster. "The Phantasmagoria." *Film History* 8:4 (1996).

Marx, Karl. *Capital: Volume One*. New York: Vintage Books, 1977.

Maverick, Lewis A. "Pierre Poivre: Eighteenth Century Explorer of Southeast Asia." *Pacific Historical Review* 10:2 (1941): 165–77.

McKendrick, Neil, John Brewer, and John Harold Plumb. *The Birth of a Consumer Society: The Commercialization of Eighteenth-Century England*. Bloomington: Indiana University Press, 1982.

McNamara, Fergal N., Andrew Randall, and Martin J. Gunthorpe. "Effects of Piperine, the Pungent Component of Black Pepper, at the Human Vanilloid Receptor (TRPV1)." *British Journal of Pharmacology* 144:6 (March 2005): 781–90.

Miller, Michael Barry. *The Bon Marché: Bourgeois Culture and the Department Store, 1869–1920*. Princeton, NJ: Princeton University Press, 1981.

Minto, Amy. "Early Insurance Mechanisms and Their Mathematical Foundations." *Montana Mathematics Enthusiast* 5:2–3 (2008): 345–56.

Mithen, Steven. *The Singing Neanderthals: The Origins of Music, Language, Mind and Body*. London: Weidenfeld & Nicholson, 2005.

Moskowitz, Marc L. "Weiqi Legends, Then and Now." *Asian Popular Culture: New, Hybrid, and Alternate Media* (2012): 1.

Mui, Hoh-Cheung, and Lorna H. Mui. *Shops and Shopkeeping in Eighteenth-Century England*. Kingston, ON: McGill-Queen's University Press, 1989.

Nauta' Lodi, Machtelt Israëls, Louis Alexander Waldman, and Guido Beltramini.

"'Rabelaiss Laughter Behind a Portrait by Holbein': Play and Culture in the Work of Johan Huizinga." *Villa i Tatti* 29 (2013).

North, Adrian C., and David J. Hargreaves. "Subjective Complexity, Familiarity, and Liking for Popular Music." *Psychomusicology: A Journal of Research in Music Cognition* 14:1–2 (1995): 77.

Ohl, John F., and Joseph Earl Arrington. "John Maelzel, Master Showman of Automata and Panoramas." *The Pennsylvania Magazine of History and Biography* 84:1 (1960): 56–92.

Oldenburg, Ray. *The Great Good Place: Café, Coffee Shops, Community Centers, Beauty Parlors, General Stores, Bars, Hangouts, and How They Get You Through the Day.* St. Paul, MN: Paragon House Publishers, 1989.

Pinker, Steven. *How the Mind Works.* New York: W. W. Norton & Company, 1999.

Poivre, Pierre. *Travels of a Philosopher; Or, Observations on the Manners and Arts of Various Nations in Africa and Asia.* Translated from the French of M. Le Poivre. T. Becket, 1769.

Potts, Daniel T. *Mesopotamian Civilization: The Material Foundations.* Ithaca, NY: Cornell University Press, 1997.

Reiss, Steven. "Expectancy Model of Fear, Anxiety, and Panic." *Clinical Psychology Review* 11:2 (1991): 141–53.

Rescorla, Robert A., and Allan R. Wagner. *Classical Conditioning: Current Research and Theory.* New York: Appleton-Century-Crofts: 1972.

Rhodes, Richard. *Hedy's Folly: The Life and Breakthrough Inventions of Hedy Lamarr, the Most Beautiful Woman in the World.* New York: Doubleday, 2011.

Riskin, Jessica. "The Defecating Duck, Or, the Ambiguous Origins of Artificial Life." *Critical Inquiry* 29:4 (2003): 599–633.

———. "Eighteenth-Century Wetware." *Representations* 83:1 (2003): 97–125.

Rosheim, Mark E. *Robot Evolution: The Development of Anthrobotics.* Hoboken, NJ: John Wiley & Sons, 1994.

Rothfels, Nigel. *Savages and Beasts: The Birth of the Modern Zoo.* Baltimore, MD: Johns Hopkins University Press, 2002.

Salimpoor, Valorie N., Mitchel Benovoy, Kevin Larcher, Alain Dagher, and Robert J. Zatorre. "Anatomically Distinct Dopamine Release During Anticipation and Experience of Peak Emotion to Music." *Nature Neuroscience* 14:2 (2011): 257–62.

Salustowicz, Rafal, and Jürgen Schmidhuber. "Probabilistic Incremental Program Evolution." *Evolutionary Computation* 5:2 (1997): 123–41.

Samir, Imad. *Allah's Automata: Artifacts of the Arab-Islamic Renaissance (800–1200).* Berlin: Hatje Cantz, 2015.

Schaffer, Simon. "Babbage's Dancer and the Impresarios of Mechanism." *Cultural Babbage: Technology, Time, and Invention* (London: Faber & Faber, 1996): 52–80.

Schivelbusch, Wolfgang. *Tastes of Paradise: A Social History of Spices, Stimulants, and Intoxicants.* New York: Pantheon Books, 1992.

Shaffer, Marjorie. *Pepper: A History of the World's Most Influential Spice.* New York: St. Martin's Griffin, 2013.

Shenk, David. *The Immortal Game: A History of Chess, or, How 32 Carved Pieces on a Board Illuminated Our Understanding of War, Art, Science, and the Human Brain.* New York: Doubleday, 2006.

Simmel, Georg. "Fashion." *American Journal of Sociology* 62:6 (May 1957): 541–58.

Sismondo, Christine. *America Walks into a Bar: A Spirited History of Taverns and Saloons, Speakeasies and Grog Shops.* Oxford University Press, 2011.

Smith, Chloe Wigston. "Callico Madams: Servants, Consumption, and the Calico Crisis." *Eighteenth-Century Life* 31:2 (2007): 29–55.

Spitzer, John, and Neal Zaslaw. *The Birth of the Orchestra: History of an Institution, 1650–1815.* New York: Oxford University Press, 2004.

Standage, Tom. *An Edible History of Humanity.* New York: Bloomsbury, 2009.

— —. *The Turk: The Life and Times of the Famous Nineteenth-Century Chess-Playing Machine.* London: Walker, 2002.

Strayer, Hope R. "From Neumes to Notes: The Evolution of Music Notation." *Musical Offerings* 4:1 (2013): 1.

Styles, John. *The Dress of the People: Everyday Fashion in Eighteenth-Century England.* New Haven, CT: Yale University Press, 2007.

Taylor, Charles. *The Literary Panorama.* London: Cox, Son, and Baylis, 1810.

Tesauro, Gerald. "TD-Gammon, a Self-Teaching Backgammon Program, Achieves Master-Level Play." *Neural Computation* 6:2 (1994): 215–19.

Thomas, Bob. *Walt Disney: An American Original.* New York: Disney Editions, 1994.

Thorp, Edward O. "Wearable Computers," 1998. Digest of Papers. Second International Symposium on. 1998.

Tortora, Phyllis G. *Dress, Fashion and Technology: From Prehistory to the Present.* New York: Bloomsbury Publishing, 2015.

Turner, Jack. *Spice: The History of a Temptation.* New York: Knopf, 2004.

Voskuhl, Adelheid. *Androids in the Enlightenment: Mechanics, Artisans, and Cultures of the Self.* University of Chicago Press, 2013.

Wallin, Nils Lennart, and Björn Merker. *The Origins of Music.* Cambridge, MA: MIT Press, 2001.

Walsh, Claire. "Shop Design and the Display of Goods in Eighteenth-Century London." *Journal of Design History* 8:3 (1995): 157–76.

Weber, Thomas P. "Alfred Russel Wallace and the Antivaccination Movement in Victorian England." *Emerging Infectious Diseases* 16:4 (2010): 664.

Whittington, E. Michael, and Douglas E. Bradley. *The Sport of Life and Death: The Mesoamerican Ballgame.* London: Thames & Hudson, 2001.

Winchester, Simon. *Atlantic: Great Sea Battles, Heroic Discoveries, Titanic Storms, and a Vast Ocean of a Million Stories.* New York: Harper, 2010.

Wing, Carlin. "The Ball: The Object of the Game." *American Journal of Play* 6:2 (2014): 288.

Wolfe, Robert M., and Lisa K. Sharp. "Anti-Vaccinationists Past and Present." *BMJ* 325.7361 (2002): 430–32.

Yafa, Stephen. *Cotton: The Biography of a Revolutionary Fiber.* New York: Penguin, 2006.

Zimmer, Carl. "How Caffeine Evolved to Help Plants Survive and Help People Wake Up." *The New York Times* (2014). http://www.nytimes.com/2014/09/science.

Zola, Émile. *Au Bonheur des Dames (The Ladies' Delight).* Trans. Robin Buss. New York: Penguin Classics, 2007.

CREDITS

2: Al-Jazarī's "Elephant Clock," facsimile taken from: Ibn al-Razzāz al-Jazarī, *Compendium on the Theory and Practice of the Mechanical Arts/al-Jāmiʿ bayn al-ʿilm wa-ʾl-ʿamal an-nāfiʿ fī ṣināʿat al-ḥiyal*, ed. Fuat Sezgin, Institute for the History of Arabic-Islamic Science, Goethe University Frankfurt, Frankfurt / Main, 2002, based on the manuscript from the Süleymanyie Library, Ayasofya 3606, 4

4, 173: Universal History Archive / Universal Images Group / Getty Images

8: Musée d'art et d'histoire, Neuchâtel, Switzerland

19: Photostock-Israel / Science Source

23, 162, 251, 253: © The Trustees of the British Museum / Art Resource, NY

27, 90, 123: Hulton Archive / Getty Images

30: Michael Nicholson / Corbis Historical / Getty Images

35: © Victoria and Albert Museum, London

40: © The Print Collector / HIP / The Image Works

50: Imagno / Hulton Archive / Getty Images

52: Grey Villet / The LIFE Picture Collection / Getty Images

59: Keystone / CNP / Getty Images

66: Photo by Hilde Jensen © University of Tübingen, Germany

75: © ZKM | Karlsruhe, photo: Harald Völkl, courtesy by UdK Berlin / Siegfried Zielinski

78: J. L. Charmet / Science Source

81: Universal Images Group / Getty Images

86: Patrick Landmann / Science Source

93: Heritage Images / Hulton Archive / Getty Images

99: George Rinhart / Corbis Historical / Getty Images

CREDITS

105: © Oram trust and Fred Wood / Courtesy of Goldsmiths Special Collections & Archives

112: British Library / Science Source

114, 120, 206: Print Collector / Hulton Archive / Getty Images

117: Photo by Guilherme Garcia, 1999

127: Florilegius / SSPL / Getty Images

133: Middle Temple Library / Science Source

140: Bildagentur-online / Universal Images Group / Getty Images

151: Kean Collection / Archive Photos / Getty Images

152, 156, 174: Science & Society Picture Library / Getty Images

168, 243: Bettmann / Getty Images

176: Earl Thiesen / Getty Images

180: Mark Kauffman / The LIFE Images Collection / Getty Images

191: De Agostini Picture Library / Bridgeman Images

197: © 2016 Thomas E Forsyth, LandlordsGame.info

204: SPL / Science Source

211: DEA / C. Novara / De Agostini / Getty Images

212: Culture Club / Hulton Archive / Getty Images

218: Courtesy of the Computer History Museum

223: Keystone / Hulton Archive / Getty Images

238: PHAS / Universal Images Group / Getty Images

256: Walters Art Museum / Bridgeman Images

261: Sheila Terry / Science Source

265: V&A Images, London / Art Resource, NY

272: Ullstein Bild / Getty Images

275: Courtesy New York City Parks Photo Archive

INDEX

Page numbers in italics indicate illustrations.

Adams, Jenny, 190
Adams, John, 241
Adler, Michael, 88
advertising
  fashion, 36–39
  London shop, *23*
  magic lantern, 168
  pianola piano, *93*
Advocate, The, 214
Alberti, Leon Battista, 160
Albius, Edmond, 129–30
Alexander, Franklin, 199
Altick, Richard, 157
American Revolution, 241–42
animation
  color, technological advancements in, 178–79
  Walt Disney, 177–81
  and the laws of physics, 177–78
  multiplane camera to show visual depth, 179–81, *180*
  overlapping motion, 178
  pencil tests as early storyboards, 178
  Snow White (film), *176*, 177–81, 184
  Steamboat Willie (film), *176*, 177
  Technicolor, 178–79
Antheil, George, 95–101
architecture
  shopping mall design, 51–53
  as a trigger for emotional responses, 42–44, 48–49
artificial intelligence
  chess as the root of, 193–94
  "curiosity reward," 281
  digital simulations that trigger emotions, 184–85
  self-learning, 280–81
  Turing Test, 227
  Watson and *Jeopardy!* 227–30
artists as toolmakers, 175–81
Au Bonheur des Dames (Zola), 43–44
auditory illusions, 158–59, 165–66
automata
  clockworks, 6–7
  Digesting Duck, 7, 79
  flute player, 76–79
  "Instrument Which Plays by Itself, The," 73–76, *75*
  lifelike simulations of individual organisms, 7, 77
  "Mechanical Turk," 14
  Writer, the, 7, *8*

Babbage, Charles
  Analytic Engine, 10
  Calculating Engine, 82
  Difference Engine, 10, 14

On the Economy of Machinery and Manufactures, 10
inspired by Merlin's Mechanical Museum, 9, 184, 284
interest in the technology of the Jacquard loom, 80–82
Baghdad (formerly Madinat al-Salam), 1–3
city design, 1–3
House of Wisdom (Bayt al-Hikma), 3
intellectual culture, 3–5
ball, importance of the, 210–15, *211*, *212*
Ballet Mécanique, 95–98
Balmat, Jacques, 263
Banu Musa, 3–5, 73–76
Banvard, John, 167, 172, 266
Barbon, Nicholas, 30
Barker, Robert, 5, 160–64, 167
baseball
Cooperstown, New York, 199–200
lineage of, 199–200
Baudrillard, Jean, 273
Beethoven, Ludwig van, 166
Bellier-Beaumont, Ferréol, 129–30
Berry, Miles, 89
Birth of A Consumer Society, The (McKendrick, Brewer, and Plumb), 37
black belt, the, 33–34
Black Cat Tavern, 242–44
Black Death, 136–37
bodily humors, 134–35
bone flutes, 65–70, *66*
Le Bon Marché, 41–46, *45*
Book of Games of Chance, The (Cardano), 205, 207
Book of Ingenious Devices, The (Banu Masu), 3–5, *4*, 73
Book of the Knowledge of Ingenious Mechanisms, The (al-Jazari), *2*, 3–5
Boorstin, Daniel, 183
Boucicaut, Aristide, *40*, 41–42, 48–49
Bradley, Milton, 195
Brand, Stewart, 219–20
Braudel, Fernand, 39–40
Brewer, John, 37
Brewster, David, 154–56, *156*, 160
Brewster Stereoscope, 160
British East India Company, 28
British Magazine, 39
British Museum, 256–57
Brunelleschi, Filippo, 160, 175, 179
brutality of the Dutch regime
Bandanese people of the Spice Islands, 119
Caribbean, *120*, 120–21
Burrows, Edward G., 234
Burton, Mary, 235

"cabinet of wonders" (*Wunderkammerns*), 255–57, *256*
caffeine
as a memory enhancer, 247–48
as a natural weapon of the coffee plant, 247
calico
"Calico Madams," 28
made popular by window displays, 31
vivid colors of chintz and, 26–27, *27*
capsaicin, 142
Cardano, Girolamo, *204*, 205, 207–209, 222
Carlyle, Thomas, 153
casino games, 221–27
Caxton, William, 188
Cecil, William, 240
celebrities, 182–84
Cessolis, Jacobus de, 187–92, 194
chance. *See* randomness
changes in society

biological changes as a result of, 48
causes of, 32, 283
phase transitions, 181–82
Charles II (king), 251–52, 259, 275
chess, *191*
allegorical power of, 192
Jacobus de Cessolis, 187–92, 194, 200
chatrang, 202
Deep Blue, 193–94
dice to speed up the game, 203
Game of Chess, The (Cessolis), 187–92
Queen Isabella, 202
pieces corresponding to real societal roles, 188–92
regional differences of the game, 200–202
worldwide appeal of, 200–203
chunking, 193
cinema. *See also* animation
as an improvement over magic lantern shows, 170–71
Carthay Circle Theatre premiere of *Snow White*, 181
close-up shot, 171
Walt Disney, 177–81
Her (film), 184
multiplane camera to show visual depth, 179–81, *180*
origin of storyboards, 178
Snow White (film), *176*, 177–81, 184
Steamboat Willie (film), *176*, 177
city planning
Fort Worth, 54
population shifts from urban centers to the suburbs, 54–55
Victor Gruen's vision, 53–55, 58–59
Walt Disney's vision for EPCOT, 55–62, *59–60*, 274
Civilization and Capitalism (Braudel), 39–40
Civil War, 34
class differences
broken down by the emerging fashion industry, 38–40
distribution of wealth as shown in the Landlord's Game, 196–98
exhibitions as great levelers, 157
public spaces as an equalizer, 246, 258–59
as shown in the game of chess, 188–90
Claude glass, *265*, 265–66
clocks as the basis for automata, 6
cloves, 111–13, 122–25, *140*
codes
cycle of encoding and decoding, 92
"talking drums" of West Africa, 91
telegraph, 91
Coen, Jan Pieterszoon, 119
coffee. *See also* coffeehouses
caffeine, 246–48
taste of, 248
utilitarian purposes of, 248
"Vertue of the COFFEE Drink" (essay), 249–50
Waghorn's, 252
Woman's Petition Against Coffee, 250–51
coffeehouses, *251*, *253*
Bedford, 252–54
differences among, 252–54
eclectic decor of Don Saltero's, 255–57
intellectual networking, 254–55, 259
John Hogarth's, 252
Lloyd's, 254
London, 254
as a news source for journalists, 254
as places of productivity and innovation, 258–59
"Proclamation for the Suppression of Coffee Houses," 251–52

coffeehouses (*cont.*)
Rawthmell's, 259
Starbucks, 274
"Turk's Head, The," 249
cognitive science
and chess, 193–94
chunking, 193
color
chintz and calico, 27, *27*
cotton, dyed, 26–27
as enhanced by a Claude glass, *265*, 265–66
trends of the mid-1700s, 37
Tyrian purple, 18–21
Columbus, Christopher, 114–15, 211–14, *212*
commodity fetishism, 153–54
Common Sense (Paine), 241
Compleat English Tradesman, The (Defoe), 24
computer technology. *See also* technology
Deep Blue, 193–94
digital simulations that trigger emotions, 184–85
Expensive Planetarium, 217–18
and games, 230–31
global collaboration, 201–202, 217–20
Hingham Institute, 215–16
IBM, 193–94, 227–28, 230, 280
"low-rent" *vs.* "high-rent" product development, 220
Minecraft, 201
networks of the early 1990s, 170
PDP-1, 215–16
for purposes of non-scientific pursuits, 219–20
Claude Shannon, 221–26, *223*
software, development of, 215–19
Spacewar! 216–20, *218*
"Spacewar: Fanatic Life and Symbolic Death Among the Computer Bums," 219–20
Edward Thorp, 221–27
Turing Test, 227
Type 20 Precision CRT, 215–16, *218*
Watson, 228–30
wearable computers, 221, 225–26
Conflagration of Moscow, The, 164–66
Conroy, David, 241
Constantine the African, 134
Cooperstown, New York, 199–200
Copland, Aaron, 97
Cortés, Hernan, 213
cotton
appealing texture of, 26–28
British East India Company, 28
"Calico Madams," 28
chintz and calico, vivid colors of, 26–27, *27*
described by John Mandeville, 26
economic fears regarding the import of, 28–29
European desire for, 29–31
importing from India, 26, 28
inventions to aid in the production of fabric, 29, *30*
slavery to produce, 34–36
Cox, James, 14
criminology
physiological causes *vs.* environmental causes, 47–48
Cristofori, Bartolomeo, 88
cultural diversity in modern times, 274–76

Darrow, Charles, 198–99
Darwin, Charles, 269–70
Das Kapital (Marx), 153–54
De Coitu (Constantine the African), 134
Defoe, Daniel, 24, 28
Dell, Michael, 216

demand
for cotton fabrics, 29–31, 34–36
"desire of Novelties," 30–31
for experiencing the world through exotic spices, 137–38
for new experiences and surprises, 61
for rubber, 214
democratizing force of fashion, 38–40
department stores
as alternatives to chapels and cathedrals, 43–44
Au Bonheur des Dames (Zola), 43–44
Le Bon Marché, 41–46, *45*
commercial profitability of wandering shoppers, 41–44
credit, extending, 44
"department-store disease," 47
haggling, elimination of, 44
influence of Aristide Boucicaut, *40*, 41–42, 48–49
origins of, 41
sensory overload and disorientation, 41–42
shoplifting, 46–49
De Smet, Pieter, 137
Devil's Milk, The (Tully), 214
Devlin, Keith, 208–209
Diamond, Jared, 141, 143
dice
astragali, 205–206, 208–209
and probability, 206–207, 209
to speed up the game of chess, 203
standardized design of, 209
Dickens, Charles, 163
Digital Revolution
artistic origins of the, 83
Spacewar! 216–20, *218*
Discourse of Trade, A (Barbon), 30
Disney, Walt
animated movies, 177–81
distaste for the environment surrounding Disneyland, 55–56
plan for a "city of tomorrow," 56–62, *59–60*
teams up with Victor Gruen, 57
Disneyland, 55–56, 273
dopamine, 281
Doritos, 109–10, 143
Doubleday, Abner, 199–200
du Saulle, Legrand, 47
Dutch domination of the Spice Islands, 119–21, *120*, 125
Dutch East India Company, 119–21

Eames, Charles, 283
Eco, Umberto, 273
Ecott, Tim, 125–26
Edwards, Daniel, 249
Eiffel, Gustave, 42
1812 Overture, 164
electronic music
input mechanisms for, 105
Moog synthesizer, 102
Daphne Oram, 102–106, *105*
Oramics Machine, 102–106, *105*
Eliot, George, 136
Elizabeth I (queen), 139–41
encyclopedias as a result of a rise in intellectual curiosity, 257–58
engineering design, 3–5, 42–43, 48–49
EPCOT (Experimental Prototype Community of Tomorrow)
anti-automobile mentality, 59–60
becomes just another theme park, 60
entirely enclosed community, 58, *59–60*
"Florida Room, The," planning site, 57
Pedshed pedestrian-only zone, 59
research for, 56
Essinger, James, 80

fame, 182–84
fashion industry. *See also* department stores
  class differences become less apparent, 38–40
  clothes as a way of feigning aristocracy, 38–39
  color and style trends of the mid-1700s, 37
  early illustrated periodicals, *35*, 36–37
  start of, 36
  trends, pace of new, 37
Fermat, Pierre de, 207–209
Firebaugh, W. C., 239–40
food. *See also* spice trade
  bioweapon strategies of plants, 142–43
  cloves, 111–13, 122–25, *140*
  cultivation of spices in new parts of the world, 121–25
  Doritos, 109–10, 143
  exploring new experiences and tastes, 143–44
  flavors in tropical locations, 142
  global nature of modern, 110, 125
  Manner of Making Coffee, Tea, and Chocolate, The, 126
  nutmeg, 113–15, *114*, 122–25
  pepper, 116–19
  spices to aid nutritional needs, 131–36
  vanilla, 125–29, *127*
Fort Worth plan, 54
Fox, John, 210
Francis, Samuel, 90
Franklin, Benjamin, 254
frequency hopping, 100–101
Freud, Sigmund, 48

Gabler, Neil, 180–81
Gama, Vasco da, 27, 31
Game of Chess, The (Cessolis), 187–92
games. *See also* dice
  artifacts from, 210
  astragali, 205–206, 208–209
  backgammon, 205–206, *206*
  ball, 210–15, *211*, *212*
  baseball, 199–200
  of chance, 205–209, 221–22
  Checkered Game of Life, 195
  chess, 188–94, *191*, 200–203
  effect on civilization, 192
  everyday metaphors taken from, 192
  Jeopardy! 228–31
  Landlord's Game, 196–98, *197*
  Mansion of Happiness, 194–95
  Minecraft, 201
  Monopoly, 196–99
  as moral instruction, 194–95, 197
  played around the world, 200–203
  roulette, 221–27
  rule-governed, 231
  Spacewar! 216–20, *218*
  ullamalitzli, 213
  video game industry, 218–20
garment design
  shift to fashion from function, 20–21
Gately, Ian, 240–41
gay bars
  Black Cat Tavern, 242–44
  Pfaff's Beer Cellar, 242–43
  as places for artists to network, 242–43
  as places of confrontation, 243–44
  Stonewall Inn, 244
gay rights movement, 244
gender issues
  employment of women in department stores, 46
  English coffeehouses as places for men only, 250, *251*

as shown in the game of chess, 189
women as shoplifters, 46–47
George, Henry, 195–96
Ginsberg, Allen, 243
Gladwell, Malcolm, 54
global collaboration
Linux and other software projects, 202
Minecraft, 201
Spacewar! 217–20
global nature of modern food, 110, 125
Gombaud, Antoine, 207
Gonod, Benoit, 89
Goodyear, Charles, 214–15
Gould, Stephen Jay, 174
Graetz, J. Martin, 217
"Grand Challenges" tradition at IBM
Deep Blue and Gary Kasparov, 193–94
Watson and *Jeopardy!* 227–30
Great Good Place, The (Oldenburg), 246
Green, Matthew, 252–55
Griffith, D. W., 171
Gruen, Victor, 49–55, *50*, 57–59
Guns, Germs, and Steel (Diamond), 141, 143
Gurk, J. J., 167

Habermas, Jürgen, 245–46, 250
Hagenbeck, Carl, 270–71, 273
Halley, Edward, 207
Hardwick, M. Jeffrey, 49
Heart of Our Cities, The (Gruen), 53–54, 57
Hegel, Georg, 151
Her (film), 184
Herrera y Tordesillas, Antonio de, 213
Horn, Paul, 227–28
horror films, precursors to, 149–50
Horsmanden, Daniel, 234–35
How the Mind Works (Pinker), 70–71
Hughson, John, 233–37, 275
Hughson's tavern
Caesar (slave), 234, 235–37
casual intermingling of races, 234–37
Cuffee (slave), 234–36
John Hughson, 233–37
"Newfoundland Irish Beauty, the" (Margaret Sorubiero), 233–34, 236
"Slave Rebellion of 1741," 234–37
human settlements, ideal climates for, 141
"hummingbird effect, the," 12, 244–45
humors, bodily, 134–35
Huygens, Christiaan, 207–208
Huygens, Lodewijk, 207

IBM, 193–94, 227–28, 230, 280
illusions
David Brewster, 154–56, *156*
distorted perception of reality, 183–85
Gespenstermacher ("ghost maker") of Leipzig, 148–50
magic lantern, 150, *152*
Phantasmagoria, the, 150–55, *151*
Paul Philidor, 149–50
proliferation of, 166–69
to re-create experiences, 163–64
Johann Georg Schröpfer, 148–49
Image, The (Boorstin), 183
Industrial Revolution
causes of, 31–32
influencers of, 32–33
Inglis, Fred, 182, 184
innovation
department stores and malls as unique destinations, 61

innovation (*cont.*)
developed at coffeehouses, 259
Disney's ideas regarding the city of the future, 55–62, *59–60*
global origins of, 12–13
"hummingbird effect, the," 12, 244–45
and music, 72
punch cards to program a loom, 80–83
intellectual culture of Baghdad, 3–5
Banu Musa, 3–5, 73–76
engineering design, books detailing, 3–5
House of Wisdom (Bayt al-Hikma), 3
intellectual curiosity
college as a time of, 257–58
encyclopedia, origin of the, 255–58
intermedi
instruments used, 84–85
origins of opera, 83–85
invention stories, misleading
baseball, 199–200
Monopoly, 198–99
rubber vulcanization, 214–15
Isabella (queen), 202
Islamic faith spread by spice traders, 113

Jacobs, Jane, 54, 61
Jacquard, Joseph-Marie, 80–83, *81*
Jaquet-Droz, Henri-Louis, 7
Jaquet-Droz, Pierre, 7
al-Jazari, Ismail, 3–5
Jefferson, Thomas, 124, 126
Jennings, Ken, 228–30
Jeopardy! 228–31
Jobs, Steve, 220
John of Eschenden, 136
Johnson, Samuel, 14
Jonze, Spike, 184
Kasparov, Gary, 194
Kay, Alan, 217
Kempelen, Wolfgang von, 14
keyboards
QWERTY, 86–87
for string instruments, 88–89
to write musical notes, 89
kleptomania, 46–49

Lacassagne, Alexandre, 47–48
Lamarr, Hedy, 98–101, *99*
Lancaster, James, 139
Land and Freedom, 196
learning methods, 280–81
leisure time, 11, 259–60
Letters on Natural Magic (Brewster), 155, 157
linear perspective, 160–61
Literary Gazette, 169
Lombroso, Cesare, 47–48
London Magazine, 39
Loth, Vincent, 119
Lovelace, Ada, 82

Macky, John, 258–59
Madinat al-Salam (now Baghdad), 1–3
Maelzel, John Nepomuk, 164–67
magic, natural, 155–56
magic lantern, 150, *152*, *168*, 171–72, *173*
Magie, Lizzie, 195–99
malls
in decline because of their predictability, 61
global development of, 53
influence of Victor Gruen, 49–55, *50*
as research for Disney's EPCOT, 56–57
Southdale Center, 49, 51–54, *52*
Mandeville, Bernard, 37
Mandeville, John, 26

Manner of Making Coffee, Tea, and Chocolate, The, 126
al-Mansur, Abu Ja'far, 1–3
al-Manum, Abu Ja'far, 3
Marguin, Jean, 77
Mártir d'Angleria, Pedro, 213
Marx, Karl, 10, 14, 153–54
McKendrick, Neil, 37
McLuhan, Marshall, 201
Medici, Ferdinando I de' (Grand Duke of Tuscany), 83–84
medicine
 and Black Death, 136–37
 compounds made by spicers, 135–36
 derived from plants, *133*
 sexual dysfunction, treatment for, 134
 spices as unproven, 136
 theriac, 135
 vanilla as, 133–34
Méliès, Georges, 171
Merlin, John-Joseph, 6, 7–10
Merlin's Mechanical Museum, 6, 8–10
Middlemarch (Eliot), 136
Miller, Michael, 42–43
Mills, Abraham, 199–200
Minecraft, 201
Mithridates VI (king), 135
Monopoly
 Charles Darrow, 198–99
 false history of, 198–99
 Landlord's Game, 196–98, *197*
 Lizzie Magie, 195–99
 Parker Brothers, 199
 tax reform, as outlined in the Landlord's Game, 195–96, *197*
Mont Blanc, 262–64
Morse, Samuel, 91
movies. *See* cinema
Mumford, Lewis, 50
Murch, Walter, 175
murex snails
 sea voyages in search of, 18–19
 source of Tyrian purple, 17–18, *19*
music
 Ballet Mécanique, 95–98
 boxes of the seventeenth and eighteenth centuries, 76
 consonance *vs.* dissonance, 68
 cultural invention *vs.* evolutionary adaptation, 69–72
 drawing wave patterns to produce, 103–106
 electronic, 101–106
 experimental, 95–98
 fourths and fifths, 67
 intermedi, 83–85
 origins of, 67
 phonograph, 94–95
 physics of intervals, 67–68
 Steven Pinker, 70–71
 pinned cylinder, 74–77
 pursuit of innovation in, 72
 Pythagorean tuning, 68
 and ratios, 68
 recorded, sharing, 106–107
 tempo, 96–97
musical instruments
 bone flutes, 65–70, *66*
 "Instrument Which Plays by Itself, The," 73–76, *75*
 of the Medici *intermedi*, 84–85
 Oramics Machine, 102–106, *105*
 panharmonicon, 166
 pianoforte, 88, 92
 player piano, 89, 92–95, *93*

natural selection, 269–70
nature as a relaxing escape, 260–66
Nossa Senhora dos Martires ("Pepper Wreck"), 115–16, *117*
"novelty bonus" when perceiving new experiences, 281, 282
nutmeg, 113–15, *114*, 122–25

Obama, Barack, 33–34
occult shows, 149–50
Oldenburg, Ray, 246
Olmsted, Frederick Law, 274
On Painting (Alberti), 160
On the Economy of Machinery and Manufactures (Babbage), 10
open-ended functionality of machines
  "Instrument Which Plays by Itself, The," *75*, 75–77
opium trade, 119
optical illusions, 155–56
  Brewster Stereoscope, 160
  Conflagration of Moscow, The, 164–66
  evolutionary adaptations that allow, 174–75
  eye as the source of most illusions, 157–58, 159
  Kanizsa triangle, 157, *158*
  Kopfermann cube, 158, *158*
  linear perspective, 160–61
  "Moving Panorama," 167
  Necker cube, 157–58, *158*, 159
  Panorama paintings, 160–64
  persistence of vision, 172, 184
  thaumatrope, 172, *174*
  zoetrope, 172
Oram, Daphne, 102–106, *105*
Orlando, Florida, 274
outdoors, the
  Albert Smith's Ascent of Mont Blanc (performance), 266
  biophilia, 260
  celebrated in art, 266
  Claude glass, *265*, 265–66
  fear of wilderness, 260
  Mont Blanc, 262–64
  mountaineering, 262–64
  national parks and wilderness preservation, 266
  nature tourism, start of, 264–65
  Horace-Bénédict de Saussure, *261*, 262–64

Paccard, Michel, 263
panharmonicon, 166
Panorama paintings, 160–64, *162*, 266
Parker Brothers, 199
Pascal, Blaise, 207–209
Patrickson, Thomas, 119
pepper
  biochemistry of, 142–43
  Cookbook (Apicus), 118
  as currency, 116
  Natural History (Pliny the Elder), 118
  Nossa Senhora dos Martires ("Pepper Wreck"), 115–16, *117*
  Queen Elizabeth I's quest to acquire, 139–41
  role in the fall of the Roman Empire, 118
perception, 159–60
Phantasmagoria, the, 150–55, *151*
phase transitions, 181–82
Phenomenology of Spirit (Hegel), 151
Philidor, Paul, 149–50
Philipsthal, Paul de, 5, 154
phonograph, 94–95
Pilon, Mary, 195
Pinker, Steven, 70–71
piperine, 142–43
play
  encourages exploration and innovation, 73, 282
  as insight into the future, 15
player piano
  concept of paying for new programming, 94
  difficulties synchronizing more than one, 96–97
  early versions, 89
  pianola, 92–95, *93*

pleasure, seeking, 71–73
"pleasuring grounds," 274–76
Pliny the Elder, 118, 142
Plumb, J. H., 37
Poivre, Pierre, 121–25, *123*
Popular Science, 179–80
Pound, Ezra, 97
Prelude, The (Wordsworth), 257
Prévost, Abbé, 22–24
Priestley, Joseph, 214, 254
probability, 206–209
probability theory, 207–209
programmability
  concept of paying for new programming, 94
  flute player automaton, 77–79
  "Instrument Which Plays by Itself, The," 75–76
  Jacquard loom, 80–83
  weaving machine for silk, 79–80
Progress and Poverty (George), 195–96
"Progress City," 57, 62
Prospect Park (Brooklyn, New York), 274–76, *275*
"public, the," as a societal force, 245–46
public gathering places
  coffeehouses, 249–59
  gay bars, 242–44
  Prospect Park, 274–76, *275*
  pubs, 240–46
  semiprivate nature of, 244–45
  taverns, 237–46
  "third place, the," 246
pubs
  and the American Revolution, 241–42
  opposition to, 240
  places to enjoy freedom of speech and action, 240–46
  political impact of, 241–42
punch cards
  in computer programming, 82
  to program a loom, 80–83
"pure aestheticism," 273

racial segregation, 234–37
randomness, 206–209, 222
ratios and music, 68
Ravizza, Giuseppe, 89
recreational time, 11, 259–60
remote-controlled torpedo using frequency hopping, 98–101
Ripley's Believe It or Not, 257
Robertson, Étienne-Gaspard, 150
Rosée, Pasqua, 249–50
Rossell, Deac, 149
roulette, 221–27
rubber
  Aztec sport of *ullamalitzli*, 213
  ball games in Hispaniola, 211–13, *212*
  exploitation of human and natural resources, 214
  Charles Goodyear, 214–15
  made into balls by Mesoamericans, 212–13
  vulcanization for durability, 214
Rutter, Brad, 228

Salter, James. *See* Saltero, Don
Saltero, Don, 255–57
Samson, Peter, 217
Samuels, Arthur, 280
Sartor Resartus (Carlyle), 153
Saussure, Horace-Bénédict de, *261*, 262–64
Schivelbusch, Wolfgang, 131–32
Schmidhuber, Jürgen, 281
Schröpfer, Johann Georg, 148–50
Scott, Walter, 155–56
sea exploration inspired by the search for murex snails, 18–19

service industry, development of the, 44–46
Shannon, Claude, 221–26, *223*
Shenk, David, 192, 201–102
shoplifting
  "department-store disease," 47
  early studies of, 47
  physiological causes *vs.* environmental causes, 47–48
  unmotivated by economic need, 47–49
  women as the culprits, 46–47
shopping. *See also* department stores; malls
  becomes a leisurely pursuit, 22, *23*
  chain stores *vs.* mom-and-pop stores, 44
  Compleat English Tradesman, The (Defoe), 24
  credit, extending, 44
  as a form of entertainment ("agreeable amusements"), 24, 32
  haggling, elimination of, 44
  shoplifting, 46–49
  store displays in the early 1700s, 22–26
Shows of London, The (Altick), 157
Simulacra and Simulation (Baudrillard), 273
Sismondo, Christine, 235
Slave Rebellion of 1741, 237
  Mary Burton, 235
  Caesar (slave), 234, 235–37
  Cuffee (slave), 234–36
  "Newfoundland Irish Beauty" (Margaret Sorubiero), 233–34, 236
slavery
  black belt, the, 33–34
  Dutch East India Company, 119–20
  fueled by the demand for cotton, 34–36
  "Slave Rebellion of 1741," 234–37
Sloane, Hans, 255–57
Smith, Albert, 266
Smith, Edward E., 216
social order
  class differences challenged by fashion, 39–40
  distribution of wealth as shown in the Landlord's Game, 196–98
  exhibitions enjoyed by all ranks, 157
  public spaces as an equalizer, 246, 258–59
  as shown in the game of chess, 188–92
Sons of Liberty, 241–42
sound. *See also* auditory illusions; music
  drawing wave patterns to produce, 103–106
  phonograph *vs.* pianola, 94–95
  re-creation of war sounds, 165–66
Southdale Center, 49, 51–54, *52*
Spacewar! 216–20, *218*
"Spacewar: Fanatic Life and Symbolic Death Among the Computer Bums" (essay), 219–20
"Spectacle Mécanique," 7
Spice: The History of a Temptation (Turner), 115
spice trade
  Black Death, responsible for spreading, 136–37
  Bourbon Island (now Réunion), 121, 124, 128–30
  cloves, 111–13, 122–25, *140*
  development of a single distribution method by Muslim traders, 112–13
  Dutch domination of the Spice Islands, 119–21, *120*, 125
  experiencing the world through exotic spices, 137–38

nutmeg, 113–15, *114*
pepper, 115–18
Poivre's plan to cultivate spices in other parts of the world, 121–25
Spice Islands, 111, *112*, 119–21
spicer's role in the royal court, 132
spreading of Islam through, 113
trading patterns, 111–12
vanilla, 125–29, *127*
worldwide importance of, 130–44
Starbucks, 274
statistics. *See* probability theory
steam engines, 29
Steele, Richard, 254
stereoscope, 160
Stonewall Inn, 244
stores. *See* shopping
Stradivarius violins, 85, *86*
Structural Transformation of the Public Sphere, The (Habermas), 245–46
Styles, John, 29–30
surprise instinct, 280–83
synchronization difficulties
frequency hopping, 100–101
pianola, 96–97, 100–101

tactile illusions, 159
Tatler, 254
taverns. *See also* Hughson's tavern
Green Dragon, 241, *243*
as inns for travelers, 239–40
as a new kind of social space, 237–38, 245
rising standards of living, 239
Roman *tabernae*, *238*, 239–40, 242
in the ruins of Pompeii, 239
technology. *See also* computer technology
computer networks of the early 1990s, 170
digital simulations that trigger emotions, 184–85
frequency hopping, 100–101
global creation, 201–202
"global village" of Minecraft, 201
as illustrated in the work of Banu Masu and al-Jazari, 3–5, *4*
multiplane camera, 179–81, *180*
music's role in developing, 91–92, 100–101
QWERTY keyboard, 86–87
textiles
"Calico Madams," 28
cotton, 26–28
East India Company, 28
economic fears regarding the import of, 28–29
French weaving industry, 79–83
inventions to aid in the production of fabric, 29, *30*
Jacquard loom, 80–83, *81*
vivid colors of chintz and calico, 26–27, *27*
theft. *See* shoplifting
theme parks
Disneyland, 55–56, 273
fantasy world of, 273
Tierpark Hagenbeck, 271–73, *272*
Thorp, Edward, 221–27
Tierpark Hagenbeck, 271–73, *272*
torpedo using frequency hopping, remote-controlled, 98–101
toys
foreshadowing the rise of mechanized labor, 11, 14–15
as illustrated in the work of Banu Masu and al-Jazari, *2*, 3–5
trading, global
Columbus's trip to the Caribbean, 114–15
Dutch East India Company, 119
Nossa Senhora dos Martires ("Pepper Wreck"), 115–16, *117*
opium, 119
Spice Islands, 111–13, 138

trading, global (*cont.*)
spices, importance of, 130–44
Venice as a central European distribution point, 118
transient receptor potential (TRP) channels, 142–43
Travels in Hyperreality (Eco), 273
Tully, John, 214
Turing, Alan, 193, 227, 280
Turing Test, 227
Turner, Jack, 115, 125, 132
Tussaud, Marie ("Madame Tussaud"), 6
2008 U.S. presidential election, 33–34
typewriters
"printing machine," 90
Remington No. 1, 90, *90*
shorthand, 89
"writing harpsichord," 89
Tyrian purple
aesthetic response to, 21
difficulty obtaining, 18
sea exploration inspired by the demand for, 18–19, 20
as a status symbol, 18, 20, 38

Unger, Johann Freidrich, 89

vanilla, 125–30, *127*, 133–34
Vaucanson, Jacques de, 7, 77–79, *78*
Vaux, Calvert, 274
"Vertue of the COFFEE Drink" (essay), 249–50
Victoria (queen), 268
visual tricks. *See* optical illusions
Votey, Edwin Scott, 92
Voyages d'un Philosophe (Poivre), 124

Wallace, Mike, 234
Walsh, Claire, 22
War and Peace (Tolstoy), 164
Warhol, Andy, 183
Watson, Thomas, 280
Watson (computer system), 228–30
Weeks, Thomas, 10
Weiditz, Christoph, 213
Welles, Orson, 177
"Wellingtons Sieg," 166
West End of London illusions, 167–70
Madame Tussaud's wax statues, 5–6
Merlin's Mechanical Museum, 6, 8–10
Panorama paintings, 5, 160–64, *162*
Phantasmagoria, the, 5, 150–55, *151*
Williams, H. L., 187–88
Williams, Raymond, 38
Williams, Tennessee, 243
Wilson, E. O., 260
Winchester, Simon, 18
Wordsworth, William, 257–58
Wright, Frank Lloyd, 53
writing machines, 87–88, 89–90
Wunderkammerns ("cabinet of wonders"), 255–57, *256*

Zimmer, Carl, 247
Zola, Émile, 43–44
zoo, the
bringing the exotic close, 267–68
Darwin's ideas about natural selection from a visit to Regents Park Zoo, 269–70
Jenny and Tommy (orangutans), 268–69
as "rational recreation," 267–68